Michael C. Münnix

Studies of Credit and Equity Markets with Concepts of Theoretical Physics

VIEWEG+TEUBNER RESEARCH

Michael C. Münnix

Studies of Credit and Equity Markets with Concepts of Theoretical Physics

With a foreword by Prof. Dr. Thomas Guhr

VIEWEG+TEUBNER RESEARCH

Bibliographic information published by the Deutsche Nationalbibliothek
The Deutsche Nationalbibliothek lists this publication in the Deutsche Nationalbibliografie; detailed bibliographic data are available in the Internet at http://dnb.d-nb.de.

Dissertation Universität Duisburg-Essen, 2011

1st Edition 2011

Editorial Office: Ute Wrasmann | Anita Wilke

Vieweg+Teubner Verlag is a brand of Springer Fachmedien.
Springer Fachmedien is part of Springer Science+Business Media.
www.viewegteubner.de

Cover design: KünkelLopka Medienentwicklung, Heidelberg
Printed on acid-free paper
Printed in Germany

ISBN 978-3-8348-1771-6

Foreword

According to a traditional, narrow definition, physics is the science exploring the laws of the non–living part of nature, while biology studies living beings. This division stems from times in which biology was still to a large extent a descriptive science whose main purpose was to order, structure and classify the observations made in the living part of nature. This changed dramatically in recent decades. Modern biology is an explaining, quantitative science employing methods which often come from chemistry and physics. Vice versa, physicists also became interested in biological questions, and the interdisciplinary field of *biophysics* emerged. A similar process is now taking place with physics, more precisely theoretical physics, and the social sciences, particularly economics. There are three driving forces: First, there is a strong tendency in economics towards working more quantitatively. Not surprisingly, it is especially strong in finance. Second, a wealth of empirical economic data became available during the last few decades. This is indispensable for theoretical physicists whose key competence is the construction of mathematical models based on empirical information. Third, complex systems moved in the focus of research in physics. Although the ultimate definition of complex systems is still debated, most researchers would agree to viewing the economy as a good example.

The new interdisciplinary field of *econophysics* attracts talented individuals from different branches of physics. However, there is one thing many of them have in common: often, they have worked in experimental physics or, at least, they analyzed experimental data. This is natural and also helpful for a field where the challenge is to lay the theoretical foundations. Dr. Münnix is an excellent example for that. He is a gifted experimentalist who worked on quantum dots before he joined my research group to start his dissertation project in econophysics. Dr. Münnix considerably strengthened the econophysics team, which then consisted of him, Dr. R. Schäfer, two Master students and myself. He took the lead in so many of our activities and produced such an impressive series of results that his thesis deserves to be published as a book. The introduction presents econophysics in a nutshell and thus makes it possible for every physicist to become familiar with the defining issues of the field. In the research part, comprising the three directions that were of particular importance in the emergence of the financial crisis of 2008–2009: Dynamics of dependencies, Epps effect and credit risk. Various original and new results are presented. It is a great pleasure for me to recommend this book to everybody interested in econophysics!

Thomas Guhr
Fakultät für Physik, Universität Duisburg–Essen

Acknowledgements

First of all, I thank my advisor Thomas Guhr for an exceptionally good research environment. His expertise and openness to discuss attracts many students and scientists, forming a group that is not only incredibly versed in theoretical physics, but also intellectually stimulating. My special thanks go to my second advisor, Rudi Schäfer. With a lot of patience he taught me much of his deep knowledge of financial markets and complex systems. His willingness to help is remarkable. Furthermore, I thank Gene Stanley for a warm welcome and the great atmosphere in his group during my research visit to Boston University. I thank Oliver Grothe for giving insight into the concept of copulae and his ambition to work interdisciplinarily.

I thank all members of our group, especially Johannes Freese, Boris Gutkin, Jens Hämmerling, Stefan Hindel, Mario Kieburg, Vladimir Osipov, Christian Recher and Thilo Schmitt for a great working climate, fruitful discussions and chats about physics and all kinds of other topics. I thank my parents for enabling me to study physics. I am very grateful to Kristina for standing by me devotedly all these years. Particularly, I thank her for falling in love with a physicist in spite of loathing physics.

Moreover, my thanks go to the Studienstiftung des Deutschen Volkes for financial support, excellent seminars and great company. I thank the Fulbright program for financing my research trip to Boston, USA.

Last but not least, I thank my friends who helped me proofreading this work, Alexander Petersen, Max Feucker, Nikolaus Plagemann and Nikola Trbović.

Parts of this work are or will be included in the publications listed in Refs. [1–7].

Michael C. Münnix

Abstract

The central topic of this work is the analysis of statistical dependencies in financial markets. In this matter, mathematical models are developed using concepts and methods from statistical physics. In particular, aspects that had a key role in the emergence of the financial crisis 2008–2009 are studied.

This work is organized in three parts. Methods are developed that both provide insight into the statistical dependence structure and significantly increase the precision in estimating correlations. As correlations have countless applications in the financial industry, this permits to enhance the estimation of risk also in existing models.

The first part introduces a similarity measure that can be used to classify typical states of a financial market and to identify their dynamics. In an empirical study, the versatility of the similarity measure is demonstrated by identifying critical states of a financial market. Moreover, in a large-scale empirical study, the structure of statistical dependencies is disclosed by the concept of copulae. Structural properties of the copula are extracted and mapped to the corresponding correlations in order to disclose the scope of correlations.

In the second part, statistical causes for the Epps effect are identified and compensation methods are developed. The Epps effect refers to the phenomenon of declining measured correlations on high frequency financial data. A major portion of the Epps effect can be compensated leading to a significant improvement in the estimation of correlations. The developed compensation methods do not require model calibrations nor is an adjustment of model parameters necessary.

In the third part, a structural model for the estimation of credit risk is discussed. By using Random Matrix Theory it is demonstrated that the existence of correlations severely limits the effect of diversification in a credit portfolio. Under the assumption of randomly fluctuating correlations, a lower bound for the estimation of credit risk is calculated.

Contents

List of Tables

List of Figures

1 Introduction

1.1 Physics and Finance

In recent years, a new trend in economic modeling has emerged. Concepts of physics are being transferred more and more frequently to problems in economics. The field of *econophysics* [8, 9] has formed as the liaison.

The methods of approach in this field are analogous to the approach of "traditional" physics: Mathematical modeling based on empirical data. The primary difference is that in econophysics, the mathematics does not describe fundamental natural laws, but instead describes statistical laws that arise from the interactions and behavior of human beings. A financial market represents such a system – with the favorable circumstance that the actions of its participants create a wealth of data.

The dramatic events of the world economic crisis that broke out in 2008 resulted in data which contain strong correlations as well as large fluctuations, or a high *volatility*, in economic terms. Deep understanding of the underlying mechanisms during the crisis will help to improve current economic models. An important task is the development of methods that describe the current market situation.

The ever-increasing availability of empirical financial data allows us to develop improved economic models. These models are fundamentally different from many traditional models in economics, which are often based on different schools. Models in econophysics are usually motivated by and evaluated on empirical data. Certainly, these models can only describe quantitative aspects. Many processes in the economy are difficult to describe quantitatively and can only be captured indirectly. Good examples are psychological or political influences. Thus, it is certainly impossible to develop a theory that describes every economic aspect involving only concepts of physics. Rather a collaboration of scientists from various fields is necessary. However, physics can improve economic models dramatically. For example, concepts of statistical physics can help to capture and predict the dynamics of credit and equity markets. One of the most important contributions on this matter is the estimation of statistical dependence. The quantification of statistical dependence allows a significant reduction of various types of financial risk.

Another benefit of econophysics is the identification of analogies between economic and physical problems. This allows applying existing theories in physics to very different problems in economics. In particular, theories of complex systems prove to be very useful. A good example in the context is Random Matrix Theory (RMT) [10, 11] which allows to predict general features of a system, such as the eigenvalue density of a correlation matrix.

However, in physics it is often possible to carry out experiments. These experiments isolate certain features of a complex system in order to test a hypothesis. In economic systems this is very difficult. For example, on one hand, it would require vast amounts of money to provoke significant reactions on a stock market that can be observed. On the other hand, even if money is not an issue, it is impossible to isolate certain features. Thus, if a reaction is provoked, it is impossible to determine if the reaction is in accordance with a model, or if it is caused by something completely new, which has not been considered yet.

A fundamental difference to traditional physics is that the underlying laws in an economic system are subject to change in time. For this reason, it is nearly impossible to develop a full theory of an economic system on a microscopic scale. There are no basic equations such as the Schrödinger equation in economics. However, rather than describing the individual participants of an economic system, we can make significant statistical statements based on the general properties of the system.

The following section 1.2 gives a brief introduction to some elementary concepts of financial markets. Section 1.3 gives some examples for successful liaisons between physics and economics. In section 1.4 the aim of this thesis is motivated and presented.

1.2 Financial Markets

A (financial) market is by its simplest definition a place or an institution where people can buy or sell certain goods. These things can be a large variety of items, such as raw materials, industrial manufactures, metals or agricultural products on a *commodity market*. Commodity markets already existed in the copper age and are believed to be the first institutionalized markets.

Nowadays, an important financial market is the stock market, or stock exchange. Stocks are small shares of a company. But why do people trade small shares of companies, and how did they come into their possession? The underlying principle is very simple. If a company needs new capital, for example, to expand into another country, it has several options. One option is to search for investors who borrow money, either for the payment of an interest rate, or for the participation in future profits of the company. But it might be difficult to find investors, if the company needs a very large amount of capital. In this case, the company can decide to sell a part of itself in order to gain capital. But it might also be very difficult to find someone who buys a large part of a company (in exchange for a lot of money). A common solution is to split the part that shall be sold into many pieces. These pieces are called *stocks*. The company can sell its stock for a predefined price on a stock exchange. Investors can buy these stocks and receive in return a small participation in future profits, called the *dividend*, and voting rights in important decisions of the company.

But why are stocks being bought and sold even after their initial sale by the company? In principle, we can think of three main reasons.

To illustrate this first reason, let us consider the following example: Investor "A" bought a stock of a company for USD 10.00. In the previous years, the company made huge losses, so he decides to sell the stock in order to buy a different stock with the profit. But nobody will pay him USD 10.00, because the probability of a good future dividend is low. But investor "B" might offer him to buy the stock for USD 5.00 because her analysis indicates that the company will continue making losses in the near future, but it will eventually recover and produce profits at some point within three years. Thus, the two investors agree on a trade at USD 5.00. This simple example illustrates a central feature of a market. No institution dictates the price of a stock (after the initial selling). The price is a result of *supply and demand*. This already gives a hint of the embedded risk in a market, because the analysis of investor "B" can of course be wrong. She might have offered a price too high, if the company does not generate profits within the next three years.

The second main reason for stocks being traded is even more speculative. Imagine that an investor knows that a company is going to introduce a new product on the next day. He thinks that this product will change the world and therefore the company's profits will increase dramatically. He wants to buy stocks of this company very quickly. Because of his huge expectations on the company's profits, he is willing to pay a price higher than the price based on the company's dividends if he can buy the stock on this day. He can also go one step further. He might think that, once the product is introduced, other investors will want to buy the stock as well and are willing to pay a price even higher than the price he has paid. Thus, he decides to sell the stock directly on the next day, when the product is introduced. In the latter case, the motivation to buy the stock is not aimed at the company's dividends anymore. It is only based on the market participants' expectations. There is also much more risk involved than in the first scenario.

A stock can be traded around the world, on multiple stock exchanges. This leads to a third main reason for stock trading. Let us say a stock is being traded in Frankfurt for the equivalent of USD 4.00 and the same stock is being traded in New York at USD 4.10. An investor can buy then the stock in Frankfurt and sell it directly in New York and make a profit of USD 0.10. This concept is called *arbitrage.* The reason for the occurrence of arbitrage is that the mechanisms described in reason one and two are performed by the participants of all stock exchanges around the world individually. This can lead to a different price in Frankfurt than in New York, because the decision to buy or sell a stock can be very subjective. The prices on different stock exchanges thus are not always synchronized. In fact, a trader that exploits arbitrage contributes to the synchronization. In principle, arbitrage is almost risk-less profit, but often there are other kinds of risk embedded. For example, the price in New York can change to USD 3.90 at the moment the investor buys the stock in Frankfurt or the EUR/USD rate can change.

These are three main reasons for stocks being traded. Of course, there are many more. For example, for *risk management*, an investor decides whether to buy or sell a stock only based on the statistical properties of the stock. We will return to this matter later in this chapter. Nowadays, a significant amount of stock market volume is traded algorithmically. However, this type of trading mainly occurs on very short timescales. For example, one of the most important factors for a hedge fund trying to exploit arbitrage is the network cable length to the stock exchange. The timescales involved are in the range of a few milliseconds.

All three scenarios described above share a common underlying principle. A trade always occurs as a result of information. This information can be anything from the knowledge of hiring a new employee in a single company, an overall market prognosis or the detection of arbitrage. With the trades that occurred due to this information, the information itself becomes *embedded* in the stock prices. A hypothetical market in which all existing information is embedded in the prices, in which all participants act perfectly rational based on identical information and in which are no hindrance factors is called an *efficient market* [12, 13]. Although the individual interpretation of information might deviate from trader to trader, the market as a whole comes to an agreement about the price of a stock. This price coincides with the "fair" price. It reflects at every point in time the value of the corresponding company. Since a deviation from the fair price offers arbitrage, the traders will exploit this immediately and the price returns to the fair price. Changes of the fair price are induced by the arrival of new information. In an efficient market the information is assumed to be completely unpredictably. Thus, the stock prices follow a random motion.

There is controversy over the concept of the fair price. For example, during the *dot-com bubble* that burst in 2002, many stock prices of Internet companies evidently deviated considerably from the fair price. However, the concept of an *efficient market* is an important assumption in many economic theories. In a certain scope, it represents a powerful model. The difficulty lies in defining this scope.

Another example of financial markets are derivative markets. On these markets, products are traded that are based on other financial products. A good example is an *option*. An option in financial context represents the right to buy or sell a stock during a certain period in the future at a predefined price. If an investor thinks that the price of a stock will increase, but he/she does not want to buy the stock itself, he/she can buy an option on this stock with the right to buy the stock in the future at a price that is lower than his expected price. The options themselves are being traded on derivative markets. This leads to the fact that the price of an option corresponds to the average market's expectation for the underlying stock during a certain period. In other words, if the stock of a company performs very well, an option to buy at a low price ("call" option) will be much more expensive than an option to sell at a low price ("put" option).

A very important derivative market is the credit market, where products based on loans, e.g. housing credits, are traded. On this market, investors try to estimate the risk of an obligor not being able to pay back the credit (this is called a *default*). In principle, profits and losses are made because some investor's estimation of risk is superior to the estimation of another investor. A systematic underestimation of the risk on the housing credit markets was one of the main causes for the financial crisis of 2008–2009.

1.2.1 Basic Concepts

We now focus on the particular example of a stock market for introducing some basic concepts and the nomenclature commonly used in econophysics literature. Let us enumerate companies with the index k. $S^{(k)}(t)$ is the price for the stock of company number k at time t. The information of the price $S^{(k)}$, however, is not continuous, because technically it only exists for the moment when someone who wants to sell this stock and someone who wants to buy the stock agree on a certain price, i.e., when a trade occurs. Strictly speaking, the price is not defined between those points of trades, but usually $S^{(k)}(t)$ refers to the *last traded price* before time t. Many stocks are so frequently traded that $S^{(k)}(t)$ can be seen as continuous in good approximation. Because of this high frequency, stocks are almost exclusively traded electronically nowadays. The concept of pricing, however, remains the same. A stock exchange (or more precisely its *clearing office*) keeps a so-called *order book*. This order book contains entries for everyone who wants to sell or buy a certain stock at a certain price and a given amount. For example, if you want to buy 100 shares of IBM at a price of USD 10.00 each, you can advise your broker to initiate the trade. He will then place a *limit-order* at the stock exchange. The term "limit" refers to the fact that you are only willing to buy the stock for USD 10.00 or below. If somebody else informs the stock exchange that he/she wants to *sell* his stocks for USD 10.00, this information will also go into the order book. Because the prices agree, the stock exchange will clear your entries in the order book and initiate the trade. As shown in Tab. 1.1, the order book has two sides. One side contains all offers of stocks to a certain price – this is called the *ask* side. The other side's entries indicate that people are willing to buy the stock at a certain price – this is the *bid* side. The highest bid and the lowest ask price are commonly referred to as *best bid* and *best ask*. Usually, there is a gap between the best bid and the best ask, called the *spread*. If this spread does not exist, i.e., if best ask and best bid are at the same price, a trade occurs, resulting in a new spread.

The limit-order is one of two fundamental ways of trading stocks. The other one is the *market-order*. Imagine that you would like to buy a stock very urgently and you do not care about the precise price (but of course you have an idea about the price because you have the information of previous prices $S^{(k)}(t)$). This corresponds to the example of the trader in the previous section, who wants to buy a stock very quickly due to his market expectation. In this situation, your broker tells the stock exchange to buy,

bid	price	ask
	USD 10.08	100
	USD 10.07	200
	USD 10.06	150
	USD 10.05	50
100	USD 9.95	
50	USD 9.94	
200	USD 9.93	
100	USD 9.92	

Table 1.1: Example of an order book for a stock.

e.g., 100 shares of IBM. Because he does not specify a certain price, this order does not go into the order book. The stock exchange rather looks for the most inexpensive offer of this stock in the order book. Consequently, the entry is cleared from the order book and the trade occurs. This process can also be split into several orders, for example if the most inexpensive offer only consists of 50 stocks.

Because of the spread, the price can jump after a market order: If the previous traded price resulted from a "buy" market-order, the last traded price is the (previous) best ask. If the next order is a "sell" market order, the price jumps from best bid to best ask with a magnitude of the spread. To avoid these artifacts in price the information of high frequency data, $S^{(k)}(t)$ sometimes represents the *midpoint* price which is the average of best bid and best ask.

An investor is usually not interested in the actual price of a stock, but in its relative price change during a given time interval Δt. This is because he/she wants to anticipate how much profit he/she can make by investing a portion of his capital in this stock. The information is given by the relative price change of a stock, which is called the arithmetic return,

$$r_{\Delta t}^{(k)}(t) = \frac{S^{(k)}(t+\Delta t) - S^{(k)}(t)}{S^{(k)}(t)} = \frac{\Delta S_{\Delta t}^{(k)}(t)}{S^{(k)}(t)} . \tag{1.1}$$

Here, t represents a dimensionless, discrete time starting at $t = 0$. On a large time horizon, the price of a stock usually follows an exponential curve. This is caused by an equilibrium between the stock market and fixed-income investments, such as fixed deposits in a bank. If an investor deposits an amount of money V_0 at a bank with an interest rate of p, the capital $V(t)$ increases exponentially,

$$V(t) = V_0(1+p)^t \ . \tag{1.2}$$

If the investor finds stocks that potentially deliver a higher profit than given by the interest rate p, he/she will take his money from the bank and put it on the stock market. On the other hand, if his stocks are not performing well compared to the interest rate the bank is offering, he/she will take his money from the stock exchange and deposit it in the bank again. This mechanism is often referred to as *global arbitrage*. Because of this mechanism, the stock prices also follow an exponential trend in general, which is illustrated by the evolution of the S&P 500 index in Fig. 1.1. To remove this trend, we can use the logarithmic difference, or logarithmic return h

$$h_{\Delta t}^{(k)}(t) = \ln\left(\frac{S^{(k)}(t+\Delta t)}{S^{(k)}(t)}\right) \ . \tag{1.3}$$

As the exponential trend is only present on large timescales, the arithmetic and logarithmic returns are nearly the same for small return intervals Δt which can be easily shown by

$$h_{\Delta t}^{(k)}(t) = \ln\left(\frac{S^{(k)}(t) + \Delta S_{\Delta t}^{(k)}(t)}{S^{(k)}(t)}\right) = \ln\left(1 + \frac{\Delta S_{\Delta t}^{(k)}(t)}{S^{(k)}(t)}\right) \tag{1.4}$$

$$\approx \frac{\Delta S_{\Delta t}^{(k)}(t)}{S^{(k)}(t)} = r_{\Delta t}^{(k)}(t) \ . \tag{1.5}$$

It depends on the actual problem, whether arithmetic or logarithmic returns should be used. In the following, we will always use arithmetic returns.

1.2.2 Financial Risk

As already mentioned above, financial investments usually come with an embedded risk. Although this seems quite intuitive, it is very difficult to identify and quantify all sources of risk. The sources of risk can be manifold. *Market risk*, for example refers to the risk that the price of a financial asset is in general unpredictable. All assumptions that an investor bases his market expectation on are either hypothetical or based on historical data. *Political risks* can also be significant, such as tax changes or bans on exports in the country that a stock's company is located. It might be very profitable for

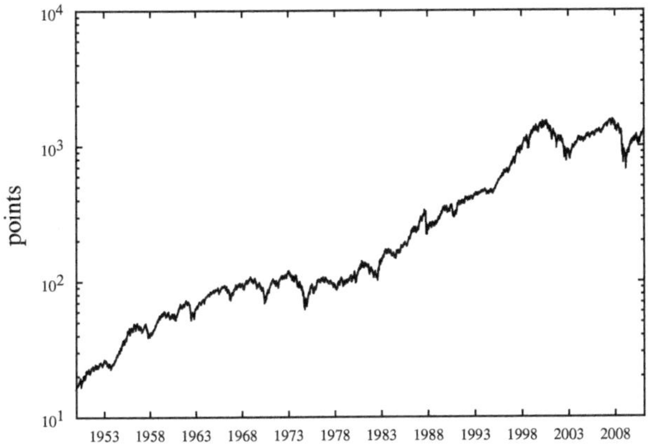

Figure 1.1: Evolution of the S&P 500 index from 1950 to 2011 on a logarithmic scale.

banks to give credit and earn the corresponding interest rates, but *credit risk*, the risk that an obligor is not able to pay back his loan is very difficult to estimate. These are just a few examples; the number of different types of risk is extensive.

However, historically these risks often induce fluctuations on the corresponding asset's returns. Hence, the standard deviation of the historical returns is commonly used to estimate and quantify the risk of an asset. The fluctuations of stock returns are referred to as *volatility* in economic terms. However, the term "volatility" does not always refer to the standard deviation, as there are various definitions of volatility in economic literature. We will use volatility as a synonym for standard deviation.

The fluctuations also include large *positive* returns, which correspond to a large profit. These contribute to the risk as much as large losses. In other words, by this definition enormous profits are not desired, because they are considered risky. An asset with a constant return, a constant interest rate, represents the asset with the lowest risk. Although the variance is only a first approximation for the quantification of financial risk, it reflects the balance between possible profit and risk. If an investor seeks to make huge profits, he/she is also exposed to large risks, i.e., the chances of large losses. On the other hand, if he/she prefers to minimize the risk, the profit will probably be small.

Terms often used in connection with risk are *covariance* and *correlation.* Certainly, the volatilities of stocks are not independent. Stocks returns can be statistically dependent, for example, if the corresponding companies operate in the same industry branch, in the same country or if they are exposed to the same risks. When quantifying the dependence of volatilities, we are only interested in the dependency of the fluctuations. Thus we need to subtract the mean values,

$$\Sigma_{kl} = \left\langle \left(r_{\Delta t}^{(k)} - \left\langle r_{\Delta t}^{(k)} \right\rangle \right) \left(r_{\Delta t}^{(l)} - \left\langle r_{\Delta t}^{(l)} \right\rangle \right) \right\rangle \tag{1.6}$$

$$= \left\langle r_{\Delta t}^{(k)} r_{\Delta t}^{(l)} \right\rangle - \left\langle r_{\Delta t}^{(k)} \right\rangle \left\langle r_{\Delta t}^{(l)} \right\rangle , \tag{1.7}$$

where $\langle \ldots \rangle$ is the average of the time series. We refer to the whole time series $r_{\Delta t}^{(k)}$ if the argument (t) is omitted. Eq. (1.7) is the well-known covariance coefficient. It quantifies the dependence of return volatilities.

Let us now consider two pairs of stocks; the first pair's stocks returns feature large variances whereas the second pair's returns have small variances. Thus, the covariance of the first pair is larger than the second pair's covariance. However, it is conceivable that the statistical dependencies of the two pairs are similar. This motivates the normalization by the standard deviations $\sigma_{\Delta t}^{(k)}$,

$$C_{kl} = \frac{\left\langle r_{\Delta t}^{(k)} r_{\Delta t}^{(l)} \right\rangle - \left\langle r_{\Delta t}^{(k)} \right\rangle \left\langle r_{\Delta t}^{(l)} \right\rangle}{\sigma_{\Delta t}^{(k)} \sigma_{\Delta t}^{(l)}} . \tag{1.8}$$

This is the Pearson correlation coefficient [14]. The value of C_{kl} gives us information about the statistical dependence of the two stocks. Its range lies between -1 and 1, where we can find the following limiting values

$$C_{kl} = \begin{cases} +1 & \text{completely correlated} \\ 0 & \text{uncorrelated} \\ -1 & \text{completely anticorrelated .} \end{cases} \tag{1.9}$$

This correlation coefficient is not independent from the choice of the return interval Δt. The length of the time series itself has also an impact, because the obtained correlations are noisier on short time series. Both aspects will be discussed in detail later in this thesis.

A good example for correlated stocks are the companies Intel and Apple. Both operate in the information technology (IT) industry branch, but they are not direct competitors. Apple uses Intel's integrated circuits in its products. Apple relies on the prices and quality of Intel's products and Intel depends on the quality and economic success of Apple's products. If something bad happens to one of the companies – let us say the brilliant chief physicist of Intel's research department dies unexpectedly, there will be consequences for Apple as well. Investors will become aware of this by the news, or see impacts on the companies' financial statements and change their expectations accordingly. This will result in a price drop of both stocks' prices – they will follow a correlated motion.

Anticorrelated stocks can be found, for example, for companies that produce seasonal products. The economic successes of the producers of ice cream and winter coats are probably anticorrelated. In a very cold winter, people will buy warm winter coats. Ice cream is mostly bought during summer. However, only few people will buy warm coats during summer or ice cream during winter. On the other hand, if two companies belong to completely different industry branches, such as Nestlé, currently the world's largest food company, and Airbus, the European aircraft manufacturer, are probably not significantly correlated.

Anticorrelations are usually very rare on the stock market. This is due to the fact that the correlation of the overall economic situation is very strong. This is often referred to as *market correlation*. Even if two companies have completely disjunct scopes of business, they might have loans from the same bank or rely on the same contractor. Moreover, if one company makes losses, it cannot pay its employees large salaries. Hence, they cannot afford to buy other companies' products. Everything is very interconnected which is why the average correlation of stock's is usually positive.

Correlations can also be used to identify stocks that belong to the same industry branch. If the correlation is large, it is very likely that the stocks belong to the same branch. The primary application of financial correlations and covariances, however, is the quantification and minimization of risk, for example in a portfolio, an ensemble of different stocks [15].

A portfolio is a linear combination of assets. These assets can be a large variety of financial products, but here we hold on to the example of a stock portfolio. The value V of a portfolio of K stocks is given by

$$V(t) = \sum_{k=1}^{K} w_k S^{(k)}(t) \ . \tag{1.10}$$

w_k defines the weight of stock k in the portfolio. It is called the *fraction of wealth* invested in stock k. The correlations within this portfolio are represented by a $K \times K$ symmetric matrix $\mathbf{C}$. Evidently, the largest values can be found on the diagonal, the correlation of a time series with itself is always one,

$$C_{kk} = 1 \quad k \in \{1 \ldots K\} \; . \tag{1.11}$$

Analogously, $\mathbf{\Sigma}$ contains all variances and covariances of the portfolio. It seems logical that the covariance matrix can be used to quantify the overall portfolio risk, which is defined as the overall portfolio variance. Hence, the portfolio risk is the weighted sum of all portfolio covariances Σ_{ij}. It is denoted by Ω^2 and reads,

$$\Omega^2 = \sum_{k=1}^{K} \sum_{l=1}^{K} w_k w_l \Sigma_{kl} = \vec{w}^\dagger \mathbf{\Sigma} \vec{w} \; , \tag{1.12}$$

where vector $\vec{w}$ contains the fractions of wealth. In this notation, the fractions of wealth need to be normalized to unity. In *portfolio optimization*, a discipline of risk management and capital allocation, one tries to minimize the portfolio risk Ω^2 with a careful choice of the fractions of wealth w_k. Put differently, an investor might decide to buy stock l, although he/she does not know much about the corresponding company itself, but because it is anticorrelated to stock k and thus lowers his portfolio risk.

Certainly, the variance of stock returns is a simple measurement for financial risk. An example of a more advanced approach is the estimation of the return that will not be undershot with a given probability α (the so-called α-quantile). The name of this concept is Value at Risk (VaR). Moreover, a general disadvantage of portfolio optimization is that the calculation of a Pearson correlation coefficient only accounts for linear statistical dependencies. Correlations give very reliable results only if the observed process can be described by a multivariate Gaussian distribution. However, as we will discuss in the next chapter, the distribution of stock returns can vastly differ from this case. Thus, the dependence of stock returns is often nonlinear and more complex. Concepts to overcome these limitations are, for example, multivariate probability distributions and copulae. However, correlations and variances are still the most commonly used techniques today. In fact, they work well in "quiet" times, but in times when the stock market gains a lot of momentum, such as during a financial crisis, they can underestimate risks dramatically.

1.3 Contributions of Physics to Economics

The key competence of every theoretical physicist is the mathematical modeling based on empirical data. The success of econophysics is rooted on this competence. This might sound trivial because mathematical modeling is the day-to-day work of many theoretical physicists. However, expertise of the careful development of a model and the subtle discussion of its scope form the basis for the success of this field. In the following, we will present some examples of contributions to economic research that originate in physics.

1.3.1 Geometric Brownian Motion

One characteristic feature of stock return time series is the abundance lot of apparently arbitrary fluctuations on small timescales, in the order of seconds to hours, but there seems to be a clear trend on larger timescales. The fluctuations on small timescales can be caused by speculative trading or arbitrage, for example. They do not correspond to the economic success of the company and thus can be seen as noise. The trend on larger timescales, however, is tied to the company's economic success.

To model this behavior, let us first turn to the arbitrary portion of this motion. The price S at time t can then be written as a sum of random price changes ΔS_n for N time steps,

$$S(T) = \sum_{n=1}^{N} \Delta S_n + S(t=0) \ . \tag{1.13}$$

As we do not include the trend yet, the average price change is zero – thus, the expected price after N time steps is the price at the first time step $S(t=0)$,

$$\langle S(T) \rangle = S(t=0) \ . \tag{1.14}$$

This seems evident, as the motion is completely erratic. The second moment of the expected price distribution at time t can be written as the sum of second moments of the price changes,

$$\left\langle S^2(T) \right\rangle = \sum_{n=1}^{N} \sum_{m=1}^{M} \langle \Delta S_n \Delta S_m \rangle \tag{1.15}$$

$$= \sum_{n=1}^{N} \left\langle \Delta S_n^2 \right\rangle + \sum_{n \neq m} \langle \Delta S_n \Delta S_m \rangle \ . \tag{1.16}$$

As the price steps are independent, the last term in Eq. (1.16) becomes zero. Moreover, because of this independence we omit the index n and write

$$\left\langle S^2(T)\right\rangle = N\left\langle \Delta S^2\right\rangle \ . \tag{1.17}$$

The total time T is proportional to the number N of time steps Δt. Hence, with $T = N\Delta t$, we obtain

$$\left\langle S^2(T)\right\rangle = \frac{\left\langle \Delta S^2\right\rangle}{\Delta t} T \ . \tag{1.18}$$

Apparently, the second moment is linear in time, which motivates the notation,

$$\Delta S = \sigma\varepsilon\sqrt{\Delta t} \ . \tag{1.19}$$

This process is called the Wiener process. Here, ε refers to a random variable that describes ΔS suitably and is normalized to unit variance. This random variable can be drawn from almost any probability density function, as long as its first and second moment exists. The standard deviation of the process is modeled by the variable σ.

We can now include the trend of the price in the process. We denote it by μ and write

$$\Delta S = \mu\Delta t + \sigma\varepsilon\sqrt{\Delta t} \ . \tag{1.20}$$

This is the Wiener process with drift. The Wiener process is the stochastic description of a phenomenon that can be observed in all kinds of processes in nature and apparently also on the stock market. This phenomenon is named *Brownian motion*, first observed by Brown in 1827 and illustrated in Fig. 1.2. The Brownian motion is characterized as *diffusive* because the second moment is linear in time, in contrast to a *ballistic* motion where this relation is quadratic. One of the first mathematical descriptions of the Brownian motion was performed in 1900 by Bachelier in his Ph.D. thesis "Théorie de la Spéculation" (The theory of speculation), which comprises a stochastic analysis of option markets and stock markets [17]. Independently, 1905 Einstein developed a mathematical description of Brownian motion. His work was an important step towards the understanding of matter composed of atoms [18]. For the first time, he provided a method that allowed experimental physicists to count atoms using ordinary microscopes. The independent works of Bachelier and Einstein on the same topic but in completely different areas – economics and physics – are a good example for the many parallels between these fields.

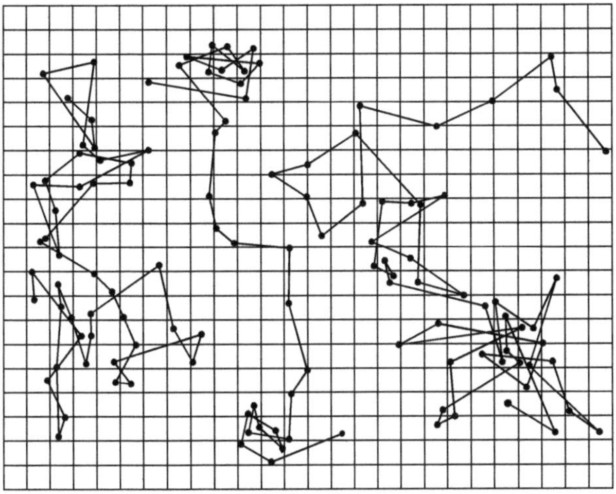

Figure 1.2: Brownian motion as measured by Perrin. The figure illustrates the movement of colliding particles with a radius of 0.53 µm. The grid size is 3.2 µm. Reproduced from [16].

The Wiener process seems suitable for the description of stock return time series, but a time series of prices has additional features. First, the price cannot be negative of course. Second, due to the global arbitrage phenomenon we discussed in section 1.2.1 stock prices grow exponentially on large time scales. Third, it is observed that the fluctuations of the price change with the amplitude of the price. These circumstances are not included in Eq. (1.20). For example, the variance is small if the price is close to zero, but if the price increases, so do its fluctuations. The magnitude of the next price change depends on the price itself. We can formalize this with

$$\Delta S = S\left(\mu \Delta t + \sigma \varepsilon \sqrt{\Delta t}\right) , \tag{1.21}$$

which is called *geometric Brownian motion*. If we draw ε from a standard normal distribution, the probability of the price $S(T)$ is given by a log-normal distribution. If we draw ε from a normal distribution, the geometric Brownian motion generates an exponential trend as well.

The geometric Brownian motion has countless applications in economic modeling. The probably most prominent one is the *Black and Scholes model*, a model for the calculation of option prices that is very widely used. The key assumption is that the price of the option's underlying stock follows a geometric Brownian motion. It was developed by Black, Scholes and Merton and first published in 1973 [19].

1.3.2 Probability Distributions

We briefly discussed that stock returns can be used to calculate correlations between stocks and thus represent a key ingredient for risk management. But the returns themselves are also of central importance because very large negative stock returns, e.g., during a financial crash, indicate dramatic events in the economy. A natural question to ask is how probable those large negative returns are, or more generally: How large is the probability density of a stock return of a certain magnitude x? The standard approach is the calculation of the probability distribution. One distinguishes two functions. First, the cumulative distribution function (cdf),

$$F_X(x) = P(X \leq x), \tag{1.22}$$

which is the probability P that the random variable X (in our case, the stock return) is smaller or equal to x. Second, the probability density function (pdf),

$$f_X(x) = \frac{dF_X(x)}{dx}. \tag{1.23}$$

Of course, $f_X(x)$ is normalized to unity,

$$\int_{-\infty}^{+\infty} f_X(x)dx = 1 \quad \text{and therefore} \tag{1.24}$$

$$\lim_{x \to \infty} F_X(x) = 1 \ . \tag{1.25}$$

The assumption of a geometric Brownian motion on an exponential trend, as discussed in the previous section, implies Gaussian distributed stock returns and log-normal distributed stock prices. As already mentioned in section 1.2.1, both is consistent with empirical data on large time-scales, e.g., daily data. Hence, the geometric Brownian motion appears to be a fair assumption in this case.

However, physicists have found that this assumption is completely unjustified for empirical data on small timescales [20, 21]. Fig. 1.3 shows the pdf

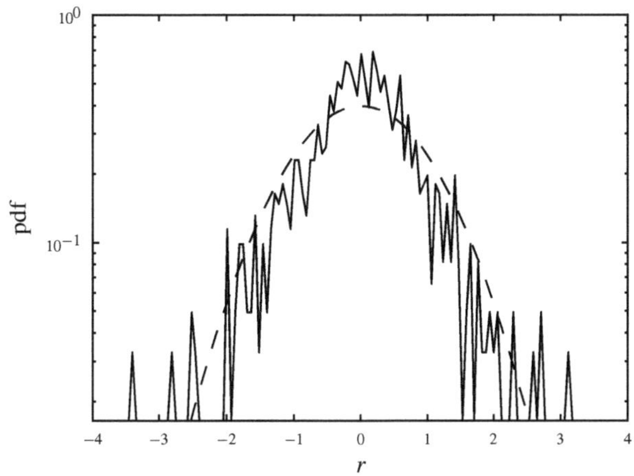

Figure 1.3: Probability density distribution of normalized daily returns of the Intel Corp. (INTC) stock from 2007 to 2010 on a logarithmic scale. The dashed curve illustrates the comparison with a standard normal distribution.

for daily returns ($\Delta t = 1\text{d}$) compared with a Gaussian distribution. For this return interval and the monitored period, the Gaussian distribution is a fair approximation. However, if we decrease the return interval down to intraday returns (Δt in the order of minutes), we obtain a distribution that differs considerably from the Gaussian case, as shown in Fig. 1.4. Here, the probability for very large and very small returns is much higher than given by the Gaussian distribution. These outer regions of the empirical distribution are often referred to as "fat tails" or "heavy tails". If assuming a Gaussian distribution, one underestimates the probability for these large fluctuations. The risk is misjudged significantly. It has been shown that the tails of the empirical cumulative return distribution follow a power-law,

$$F(r) \propto r^{-\alpha} . \tag{1.26}$$

The parameter α is approximately 3, but this also changes with the size of the return [20, 21]. Because of these algebraic tails, it is difficult to find a suitable probability density function. There are several candidates. A promising one is the Lévy stable distribution. However, simply the knowledge that one underestimates the risk if assuming Gaussian distributed returns disclosed a fundamental weakness of financial modeling that is still commonly encountered in practice.

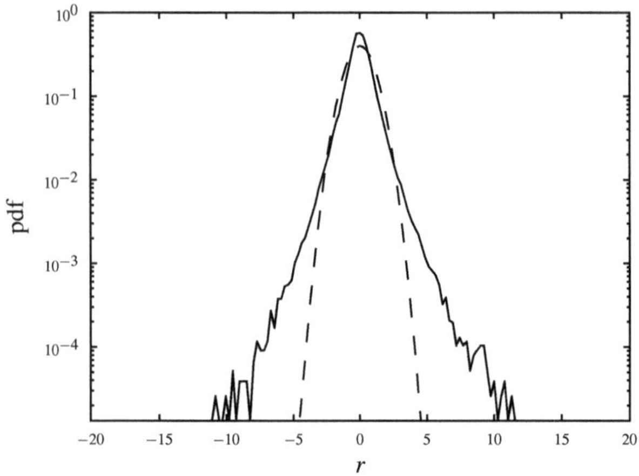

Figure 1.4: Probability density distribution of normalized 5-min returns of the Intel Corp. (INTC) stock from 2007 to 2010 on a logarithmic scale. The dashed curve illustrates the comparison with a standard normal distribution.

Moreover a short-come of many traditional economic models is the assumption of a constant volatility. Empirical studies have shown that the volatility usually fluctuates in time. Periods of large volatiles tend to appear in clusters in time [22, 23]. This led to the development of GARCH models, which we will discuss in chapter 3.

1.3.3 Random Matrix Theory

Random Matrix Theory (RMT) is a good example for common concepts of physics and finance. Originally it was introduced by Wigner in 1967 to deal with the statistics of eigenvalues and eigenfunctions of complex many-body quantum systems. In addition to many applications in physics it also is very useful for investors. The ansatz of RMT is the following. One exchanges the Hamiltonian of a system with a matrix that has completely random entries, but shares certain properties with the Hamiltonian, i.e., symmetry or invariance against transformations. Despite the randomness, one can make profound statements about the system's statistical properties. The motivation of this approach is that in complex systems, the Hamiltonian is often unknown, but certain properties of it are known.

The aim is to describe the general behavior of the system rather than the microscopic processes. RMT allows us to calculate the statistical features of the system such as its eigenvalue density. This approach is similar to the ergodicity in thermodynamics, where one can make statements about the system, e.g., if time goes to infinity (or the time span is very long). Ergodicity in RMT increases the number N of random Hamiltonians to infinity to make statements about the system's statistics.

In RMT, several ensembles exist, consisting of a symmetry and a probability density function for the entries of the Hamiltonian or the Dirac operator. These represent different systems in physics and exhibit different eigenvalue densities. In order to compare a measured eigenvalue density the characteristic scales of the system need to be removed from the measured eigenvalue spectrum. This procedure is called *unfolding.* In our case, we want to make statements about the correlation matrix of a portfolio.

In economics there is certainly no Hamiltonian that describes, for example, the movement of stock prices. However, it is possible to formally map the correlation matrix $\mathbf{C}$ of K stocks to a Dirac operator in RMT by expressing it as

$$\mathbf{C} = \frac{1}{T}\mathbf{M}\mathbf{M}^\dagger \ , \tag{1.27}$$

where T is the length of the underlying time series and $\mathbf{M}$ is a $K \times T$ matrix. A detailed explanation for this notation is given in chapter 4. The Dirac operator $\mathcal{D}$ in the *chiral Gaussian orthogonal ensemble (chiral GOE)* in RMT is given by

$$\mathcal{D} = \begin{bmatrix} 0 & \mathbf{W}^\dagger \\ \mathbf{W} & 0 \end{bmatrix} , \tag{1.28}$$

where $\mathbf{W}$ is a random matrix. If we choose

$$\mathbf{W} = \frac{1}{\sqrt{T}}\mathbf{M} \ , \tag{1.29}$$

then $\mathcal{D}$ and $\mathbf{C}$ have identical spectra (apart from the zero modes). However, we emphasize that this is only a formal analogy. This Dirac operator does not correspond to laws of motions for the stock market.

But why is this correspondence useful for us? Financial correlation matrices are noisy. This can simply be due to the finite lengths of the time series. But how large is the portion of this noise? Clearly, if we write the correlation matrix as in Eq. (1.27) and choose $\mathbf{M}$ by Eq. (1.29), the portion of randomness would be one hundred percent. As we can calculate the

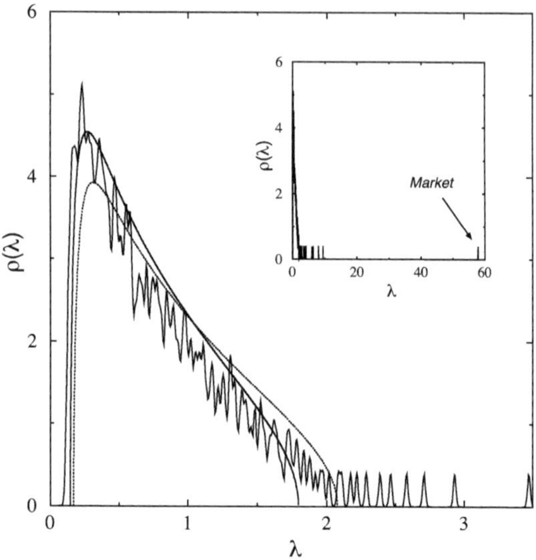

Figure 1.5: Smoothed density of the eigenvalues of the correlation matrix **C**. For comparison the density of chiral GOE for $\sigma^2 = 0.85$ (dotted lone) and $\sigma^2 = 0.74$ (solid line) is plotted. Taken from [11].

eigenvalue density spectrum of the random matrix using RMT, we can compare it with the eigenvalue spectrum of the correlation matrix and thereby *identify* the noise.

The eigenvalue density of the random matrix ensemble can be calculated analytically [24, 25] as

$$\rho(\lambda) = \frac{Q}{2\pi\sigma^2}\frac{\sqrt{(\lambda_{\text{max}} - \lambda)(\lambda_{\text{min}} - \lambda)}}{\lambda} , \tag{1.30}$$

where Q is the ratio of the length of the time series T of returns that the correlation matrix is based on to the number of stocks K,

$$Q = \frac{T}{K} \geq 1 . \tag{1.31}$$

$Q \geq 1$ is a requirement for the correlation matrix to be non-singular. σ^2 is the variance of the matrix $\mathbf{M}$'s entries. By proper normalization, we choose $\sigma^2 = 1$. λ_{min} and λ_{max} are the boundaries of the spectrum and are given by

$$\lambda_{\text{min}}^{\text{max}} = \sigma^2\left(1 + 1/Q \pm 2\sqrt{1/Q}\right) . \tag{1.32}$$

The results of an empirical analysis are shown in Fig. 1.5. Here, the correlation matrix **C** is calculated from 406 assets of the S&P 500 during the years 1991-1996. When comparing the spectra, an immediate observation is that the largest eigenvalue λ_1 is 25 times larger than the predicted $\lambda_{\max}$. The corresponding eigenvector represents the "market" itself. It has approximately equal components on all of the K stocks. A first assumption is that the components of the correlation matrix, which are orthogonal to the market are pure noise. This amounts to subtracting the contribution of $\lambda_{\max}$ from the nominal value $\sigma^2 = 1$, leading to $\sigma^2 = 1 - \lambda_{\max}/N = 0.85$. This gives the theoretical distribution for a matrix that is purely random except for its largest eigenvalue. An improved agreement with the bulk of the density distribution can be obtained with a smaller value of $\sigma^2 = 0.74$, corresponding to 94% of the spectrum.

Thus we are able not only to estimate the amount of noise in the correlation matrix but also to *identify* it in the eigenvalue density spectrum. A common practice is to transform the correlation matrix into the diagonal form, calculate $\lambda_{\min}$ and $\lambda_{\max}$ and then "eliminate" the noise by setting the eigenvalues that correspond to noise to zero. The correlation matrix can then be transformed back and one obtains a noise-filtered correlation matrix. Although this technique might seem a bit primitive, it can significantly reduce portfolio risk [10] and is widely used in the financial industry. However, this is only one method to lower the noise in financial correlation matrices. There is a wide range of noise reduction techniques, see, e.g., Refs. [26–29].

1.4 Motivation and Outline

The previous sections underline the importance of correlation – or in general, statistical dependence for the proper estimation of risk. This work deals with central problems related to statistical dependence and risk estimation. A particular focus lies on the financial crisis of 2008–2009.

What matters in risk management is to uncover statistical laws governing the time dependence so that one can anticipate periods during which the market is at risk of sudden change. At first sight, this would seem impossible since everyone knows that the graphs of financial variables as a function of time are highly irregular and at first sight unfeasible to predict. A financial market can be seen as a non-stationary system. One can wonder if physics can offer any hints of how to make progress.

In physics, one focuses on "state variables" that correspond to – and hence identify – the state of the system of interest. In finance, the analog of a state variable does not yet exist. In chapter 2, we introduce a similarity measure that enables us to compare a market's correlation structure between two different points in time. Using this measure we can identify states of the market, which correspond to different correlation structures in the financial market. As a practical example for an application we utilize the measure in portfolio optimization by calculating correlation matrices that are weighted by the market similarity. Moreover, we analyze the statistical dependence of a stock market using copulae, a method that captures statistical dependence much more precisely than correlations. The use of correlation coefficients and the corresponding Gaussian copula is discussed in relation to the financial crisis of 2008–2009 [30]. For example, using the Gaussian copula, one can vastly underestimate the probability of correlated extreme events. In a large-scale empirical study we disclose the degree of error that is involved in the Gaussian copula.

However, the estimation of copulae requires large amounts of data. This is due to the fact that we have to estimate a two dimensional function instead of a single correlation coefficient. Estimating the average pairwise copula of a whole market might be feasible, because the amount of data is large. For two single stocks on a short time horizon, the amount of data available is much smaller and thus the estimate of their copula is probably very noisy. In this case, a traditional correlation coefficient, e.g., the Pearson coefficient, is an adequate measure. Moreover, when estimating the statistical dependence for K stocks, the corresponding K-dimensional copula is considerably complex and difficult to handle analytically. In many cases, such as portfolio optimization, a $K \times K$ correlation matrix is more convenient.

Hence, we approach the topic of estimating correlations in chapter 3 where we develop methods to enhance the precision of estimating correlations in financial data. As correlation coefficients are very widely used, this has many applications and is possibly not restricted to financial data. In particular, we approach a central problem when calculating financial correlations – the dependence on the return interval Δt. For the proper estimation of financial risk, it is indispensable to calculate financial correlations as recent as possible, i.e., using short time horizons and small return intervals. However, financial correlations decline on small return intervals. This phenomenon is referred to the *Epps effect*. This effect causes distorted correlation coefficients and is a major problem in the precise estimation of correlations.

Since its discovery in 1979, many explanations have been developed that are partially incompatible with each other. As there are certainly many mechanisms contributing to the Epps effect – and most of them cannot be estimated without making considerable model assumptions, we pursue an alternative approach. We seek for causes that are purely of statistical origin and hence can be compensated without the requirement of model calibrations or model parameters.

We discuss two causes for the Epps effect that are purely of statistical origin. Moreover, we develop compensation methods that allow compensating for these causes. The compensation methods are demonstrated in a model set up. In an empirical study we quantify the contribution of these causes on the Epps effect under certain conditions.

In chapter 4, the estimation of credit risk, another topic in which correlations play a central role is discussed. The estimation of credit risk is much more complex than the risks in other fields of the economy. This is because the shape of the loss distribution is asymmetric due to the character of a *default*, which occurs if credit is not paid back. Traditional risk measures, such as the volatility of assets are not applicable to credit risk. The estimation of the loss distribution is the key requirement in the estimation the risk embedded in a portfolio of credits.

The misjudgment of credit risk was one of the main causes for the finical crisis of 2008–2009. These dramatic events emphasize the importance of a precise estimation of credit risk. We develop an analytical model to estimate the risk within a portfolio of credits. This model can be characterized as a *structural* credit risk model, i.e., we use a high level of abstraction. This allows us to gain insight into the processes in a credit portfolio. However, despite the level of abstraction, the model has a direct practical application. This application is given by a portfolio of speculative margin loans, credits that are given to investors based on their portfolio of stocks or other assets. The results can be used to estimate the lower bound of risk in a credit portfolio.

In our model, we assume random correlations with average correlation level zero. We analyze how the exposure of a credit portfolio to losses is altered by the existence of correlations and their fluctuations. The results can be used to estimate the lower bound of risk in a credit portfolio. A previous study indicates by numerical simulations that the existence of correlations destroys the effect of diversification, the reduction of risk in large portfolios [31]. We discuss this phenomenon analytically.

2 Dynamics of Statistical Dependencies in Financial Markets

In this chapter, we pursue two different approaches to give insight into the statistical mechanics of a financial market[1].

In section 2.1 we introduce a similarity measure that enables us to compare the structure of a market at two different points in time. The similarity measure allows identifying unique events such as financial crises. Moreover, we are able to identify typical states that the market adopts.

In section 2.2, we perform a large-scale empirical study to disclose the average dependency structure using copulae. We demonstrate the degree of error that is involved in the commonly used Gaussian copula and map structural features of the empirical copula to the markets' average correlation level.

2.1 Identifying States of a Financial Market

Financial markets are changing more and more rapidly nowadays. This non-stationarity can lead to a fatal misinterpretation of the involved risks. Capturing the dynamics of a financial market is an essential task to judge the current situation. We introduce a measure, based on correlation matrices, that serves to quantify the similarity of market states at two different points in time. Analyzing the S&P 500 stocks in the 19-year period 1992–2010 allows us to identify points of drastic change in the correlation structure and map these points to occurrences of financial crises. We find that a wide variety of characteristic correlation structure patterns exist in the observation time window, and that these characteristic correlation structure patterns can be classified into several typical market states. We thereby offer a method for recognizing transitions between different market states. Hence, this similarity measure can give an indication of drastic events before they are fully developed, and thus offers one the opportunity for a timely reaction.

---

[1]For details see Refs. [3, 5, 6].

This section is organized as follows. The similarity measure is introduced in section 2.1.1. In section 2.1.2, an empirical study is presented in which several states of the US stock market can be identified. As a possible application, we discuss in section 2.1.3, a portfolio optimization that utilizes the similarity measure to calculate similarity-weighted correlation matrices. We summarize the results in section 2.1.4.

2.1.1 Similarity Measure

The effort to understand the dynamics in financial markets is attracting physicists as well as economists [8, 32–38]. Statistical dependencies between stocks are of particular interest, because they play a major role in the estimation of financial risk. Since the market itself is subject to continuous change, the statistical dependencies also change in time. For example, changes in supply and demand can lead to diverse market states [39]. But how similar is the market state right now, compared to previous states? To calculate this *similarity* we measure temporal changes in the statistical dependence between stock returns.

As already discussed in section 1.2.2, a common measure for statistical dependency between two stock return time series $r^{(i)}$ and $r^{(j)}$ is the Pearson correlation coefficient C_{ij}. For the sake of simplicity, we omit the index for the return interval Δt in the notation of this chapter. When calculating the correlation coefficients of K stocks, we obtain the $K \times K$ correlation matrix $\mathbf{C}$, which gives an insight into the statistical dependencies between the stocks.

One problem we encounter in extracting useful information from empirical data is that we seek a correlation matrix from very recent data, in order to provide a good description of the current correlation structure. This is because correlations change dynamically, making it difficult to estimate them precisely [27, 40–42]. However, if the length T of the time series is short, the correlation matrices $\mathbf{C}$ are noisy [10, 11, 26, 27]. On the other hand, to keep the estimation error low, T can be increased, but this leads to a correlation matrix that may not describe the present state well.

For $T/K < 1$ the correlation matrix becomes singular. However, one can still make significant statistical statements, e.g., for the average correlation level whose estimation error decreases as $1/K$. Here we discuss a measure based on the average of the *difference* of two correlation matrices $\mathbf{C}^{(L)}(t_1)$ and $\mathbf{C}^{(L)}(t_2)$ at different times t_1 and t_2 on the time window L. For example, the correlation matrix $\mathbf{C}^{(L)}(t_1)$ is based on time series from the interval

$[t_1 - L, t_1]$. We define

$$\zeta^{(L)}(t_1, t_2) = \left\langle \left| C_{ij}^{(L)}(t_1) - C_{ij}^{(L)}(t_2) \right| \right\rangle_{ij} \tag{2.1}$$

to quantify the difference of the correlation structure for two points in time, where $|\cdot|$ denotes the absolute value and $\langle \ldots \rangle_{ij}$ denotes the average over all components. Because we study only the average difference, $\mathbf{C}^{(L)}(t_1)$ and $\mathbf{C}^{(L)}(t_2)$ can be calculated on short time windows.

2.1.2 Empirical Study

To apply the above general statements to a specific example, we analyze two datasets: (i) we calculate $\zeta^{(L)}(t_1, t_2)$ based on the daily returns of those S&P 500 stocks that remained part of the S&P during the 19-year period 1992–2010, and (ii) we study the four-year period 2007–2010 in more detail based on intraday data from the NYSE TAQ database [43]. Since the noise increases for very high-frequency data (see chapter 3), we extract one-hour returns for dataset (ii). For one-hour returns, we consider this market microstructure noise as reasonably weak.

However, sudden changes in drift and volatility are present on all time scales. They can result in erroneous correlation estimates. To address this problem, we employ a local normalization method by Schäfer and Guhr [44] on dataset (i). For each return $r(t)$ we subtract the local mean and divide by the local standard deviation,

$$g_{\text{loc}}(t) = \frac{r(t) - \langle r(t) \rangle_n}{\sqrt{\langle r^2(t) \rangle_n - \langle r(t) \rangle_n^2}} \,. \tag{2.2}$$

The local average $\langle \ldots \rangle_n$ runs over last n most recent days. $n = 13$ yields nearly normal distributed time series.

We calculate the correlation matrices of dataset (i) for disjunct two months windows. This corresponds to $L = 32$ daily stock returns. The results are presented in Fig. 2.1a. This representation gives a complete overview about large structural changes of this financial market of the past 19 years in a single figure. It allows comparing the similarity of the market states at different times. To make this procedure concrete, consider the following example. Pick a point on the diagonal of Fig. 2.1a and designate it as "now". From this point the similarity to previous times can be found on the vertical line above this point, or the horizontal line to the left of this point.

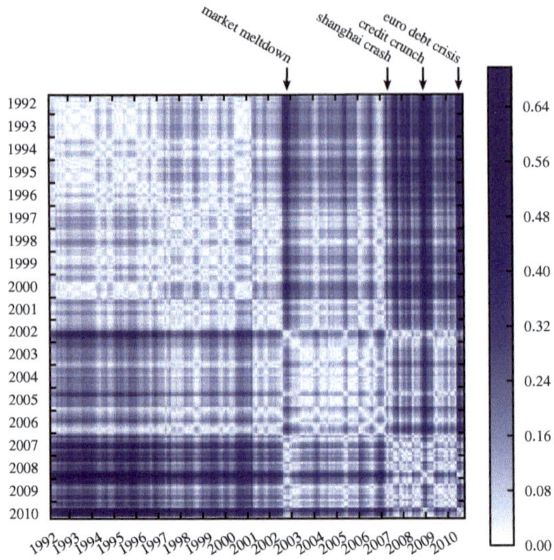

(a) daily data

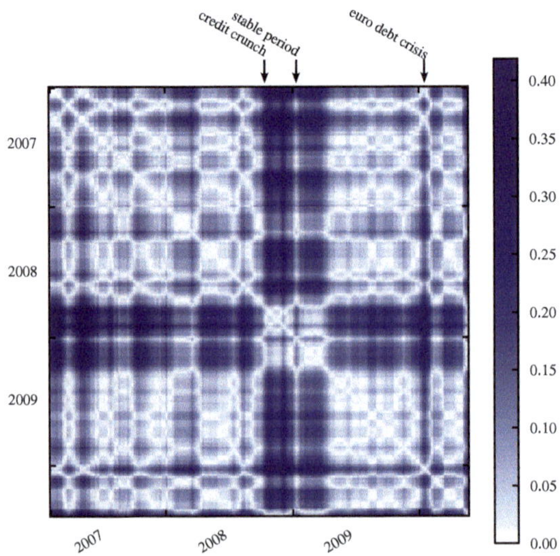

(b) intraday data

Figure 2.1: The market similarity $\zeta^{(L)}$ for two different timescales. in Fig. (a) is based on daily data with $L = 32$. Fig. (b) is a more detailed study of the 2007–2010 period using intraday data and $L = 4$. The area of Fig. (b) corresponds to a magnification of the lower right square in Fig. (a).

Light shading denotes similar market states and dark shading denotes dissimilar states. We can furthermore identify times of financial crises with dark shaded areas. This indicates that the correlation structure completely changes during a crisis. There are also similarities between crises, as between the "credit crunch" that induced the 2008–2009 financial crisis and the "market meltdown", the burst of the dot-com bubble in 2002. A further example is the overall rise in correlation level in the beginning of 2007. This event can be mapped to drastic events on the Shanghai stock exchange [45].

Using dataset (ii) we are able to obtain a more detailed insight into recent market changes, as shown in Fig. 2.1b. This area is represented by the lower right square in Fig. 2.1a. Using intraday data allows us to calculate the correlation matrices on shorter time scales. We choose a time horizon of one week (4 daily returns, $L = 4$), because it provides insight into changes in the correlation structure on a much finer time scale, enables us to identify a short sub-period within "credit crunch" (in the beginning of 2009) during which the market temporarily stabilizes before it returns to the crisis state. This phenomenon might be related to the market's reaction to news about the progress in rescuing the American International Group (A.I.G.) [46, 47].

The evolutionary structure presented in Figs. 2.1a and 2.1b illustrate that the correlation matrix sometimes maintains its structure for a long time (bright regions), sometimes changes abruptly (sharp blue stripes), and sometimes returns to a structure resembling a structure the market has experienced before (white stripes). This suggests that the market might move among several typical market states. To extract such typical market states, we perform a clustering analysis of the results of dataset (i) [2].

This clustering analysis is based on a *top-down* scheme: All the correlation matrices are initially regarded as a single cluster and then divided into two clusters by the procedure based on the k-means algorithm [48–50]. Each division step consists of the following process:

1. Choose two initial cluster centers from all matrices. Label all other matrices by the more similar cluster center, in terms of $\zeta^{(L)}$.
   a) Recast two new cluster centers to the "center of mass".
   b) Re-label all matrices to their most similar cluster center.
   c) Repeat this process until there is no change in labeling.
2. Take the best division out of all possible initial choices, which gives the least $\langle(\zeta^{(L)})^2\rangle$.

[2] The clustering analysis has been carried out by T. Shimada; See Ref. [6].

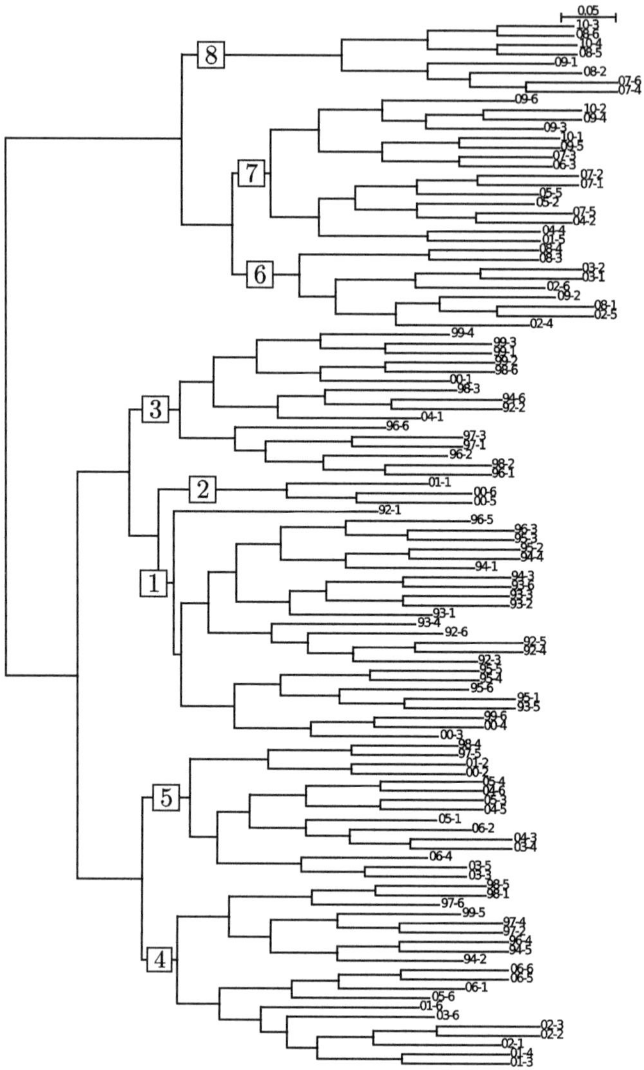

Figure 2.2: The complete similarity tree of the clustering analysis with zero threshold. Each right end of the tree corresponds to each 2-month term (year-term). Terms 1, 2, ..., 6 correspond to January-February, March-April, ..., November-December.. The horizontal length of each branch represents the distance from the center of the sub-cluster to the center of the original cluster before the last dual division.

Symbol	Industry branch
E	Energy
M	Materials
I	Industrials
CD	Consumer Discretionary
CS	Consumer Staples
H	Health Care
F	Financials
IT	Information Technology
C	Communication
U	Utilities

Table 2.1: GICS classifications.

We stop this division process when the average distance from each cluster center to its members becomes smaller than a certain threshold. To identify the *typical* market states, we choose the threshold at 0.1465, as it represents approximately the best ratio between the distance between clusters and their intrinsic radius and in the metric induced by the similarity measure.

One can obtain finer structures by choosing smaller threshold values, ultimately until all the matrices are identified as different components, as presented in Fig. 2.2. Here, no termination of the division process takes place until all the correlation matrices are identified as different components. In other words, we set the aforementioned threshold to zero. We are able to identify 8 typical market states using a threshold of 0.1465. We enumerate these states from 1 to 8 and indicate them in Fig. 2.2 as well.

The results of the clustering analysis indicate that there are "hidden" states sparsely embedded in time, in addition to regimes that dominate the market during a continuous period and thus are easily found by eye. Because of the window length of two month ($L = 32$), some smaller financial crashes cannot be resolved. Our aim is rather to identify the general evolution of the market, which is, in some cases, induced by a financial crisis.

To visualize the characteristic structures of each state, we calculate its average correlation matrix and sort the companies according to their industry branch, as defined by the Global Industry Classification Standard (GICS) [51], as listed in Tab. 2.1. The resulting matrices, the industry branches correspond to the blocks on the diagonal. The correlation be-

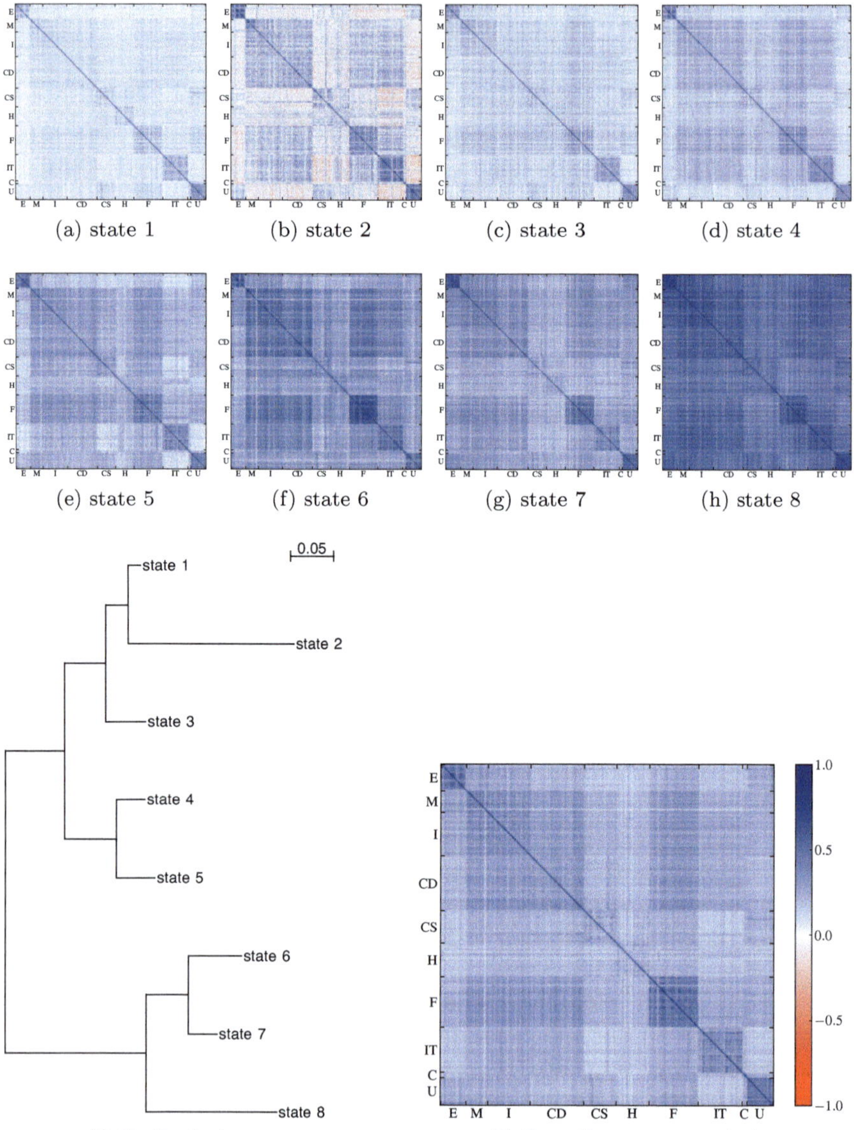

(a) state 1 (b) state 2 (c) state 3 (d) state 4

(e) state 5 (f) state 6 (g) state 7 (h) state 8

(i) Similarity tree (j) Overall average correlation

Figure 2.3: Correlation structure of different market states (a-h). Simplified similarity tree structure of the 8 market states (i) . The distance of each state to the average is illustrated by their horizontal distance. Fig (j) shows the overall average correlation matrix.

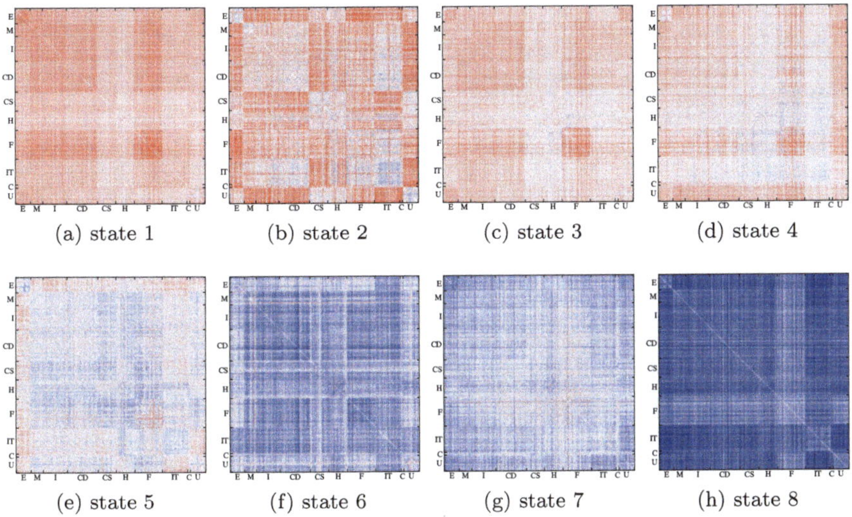

(a) state 1 (b) state 2 (c) state 3 (d) state 4

(e) state 5 (f) state 6 (g) state 7 (h) state 8

Figure 2.4: Difference matrices for the market states in figures 2.3a-h to the average correlation matrix shown in Fig. 2.3j. Illustrated using the same colormap as in Fig. 2.3j.

tween two branches is given by the off-diagonal blocks. The results are illustrated in Fig. 2.3. We can confirm that the typical states obtained from the clustering analysis indeed correspond to different characteristic correlation structures. We can see differences between the states in the correlation between branches as well as in the correlation within a branch. The correlation within the energy, information technology, and utilities branches is very strong in all states. State 1 shows an overall weak correlation, while states 3 and 4 feature in addition a strong correlation of the finance branch to other branches. State 2 shows very unusual behavior: In the period of the dot-com bubble, many branches are anticorrelated with one another. In states 5, 6 and 7, the overall correlation level rises, although certain branches, such as energy, consumer staples, and utilities, are either strongly or weakly correlated with other branches.

Some of the correlation structures in Fig. 2.3 look quite similar at first sight. Their distinctiveness can be emphasized by calculating the difference to the average correlation level. This is illustrated in Fig. 2.4. For example, state 3 and state 4 appear to be very similar in Fig. 2.3. However, Fig. 2.4 unveils that the correlation within the Energy branch (denoted with "E") is significantly different.

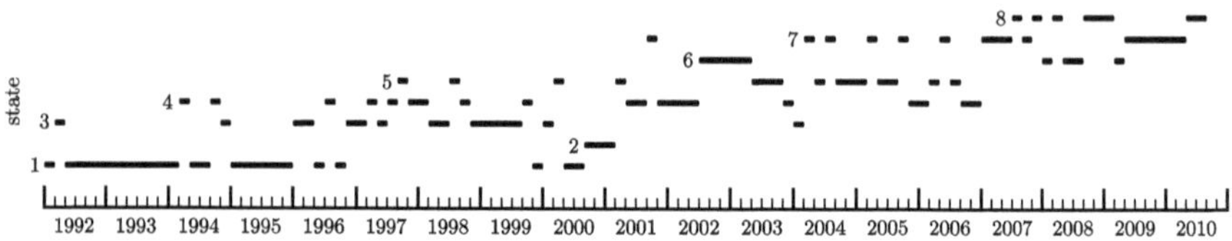

Figure 2.5: Evolution of the market state between 1992 and 2010. The states are numbered from 1 to 8.

Our analysis also offers insight into market structure dynamics. Fig. 2.5 shows the temporal behavior of the market state. The market sometimes remains in the same state for a long time, and sometimes stays only for a short time. The typical duration depends upon the state: Some states (e.g., state 1 and state 2) appear in clusters in time while other states appear more sparsely in time (e.g., state 4). There seems to exist a global trend on a long time scale, although the market state is switching back and forth between states.

Another interesting observation in Figs. 2.3 and 2.4 is that the energy branch can be either strongly correlated to the rest of the market, weakly correlated, or even anti-correlated. Therefore we study the histogram of the correlation coefficients $C_{ij}^{(32)}(t)$. We present the results in Fig. 2.6. In the months leading up to the crisis that begun in October 2008, we observe a bimodal structure in the histogram.

It corresponds to the time period when the Energy branch shows a strong anticorrelation with other branches. The bimodality suggests that a subset of stocks – in this case, predominantly the Energy stocks – decouples from the rest of the market. During the crash, the histogram shows a very narrow distribution around large values of the correlation coefficients, which corresponds to state 8 in Fig. 2.3, where the branch structure is lost almost completely in an overall strongly correlated market.

The GICS sorting enables us also to take a closer look on the sort stable period within the crisis, we identified in the beginning of this section in Fig. 2.1b. A detailed look of the correlation structure is illustrated using dataset (ii) in Fig. 2.7. While the correlation structure during the crisis displays an overall high correlation level, the correlation structure of the stable period is similar to state 7, one of the typical states in a calm period, which is identified from dataset (i).

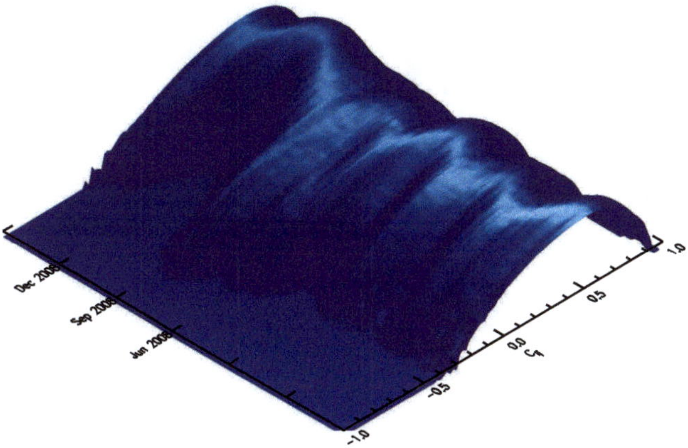

(a) Surface plot

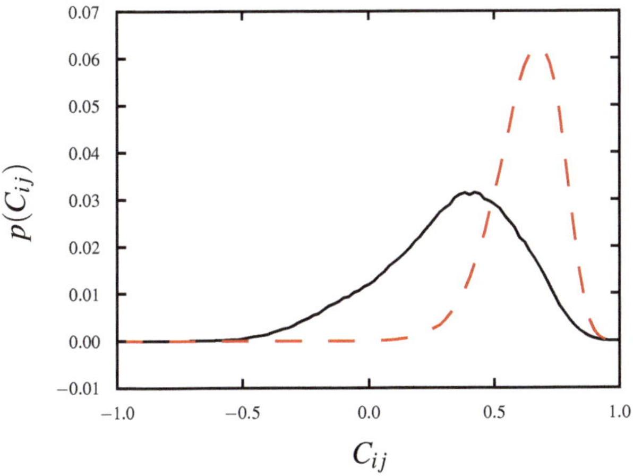

(b) Separate histograms

Figure 2.6: Footprint of the state transition in the 2008 crisis by histograms of the correlation coefficients $C_{ij}^{(32)}(t)$. (a) Surface plot for the time period September 2007 to March 2009. We use a logarithmic scale to show the bimodal structure more clearly. (b) Histograms for September 2008 (black solid line) and December 2008 (red dashed line).

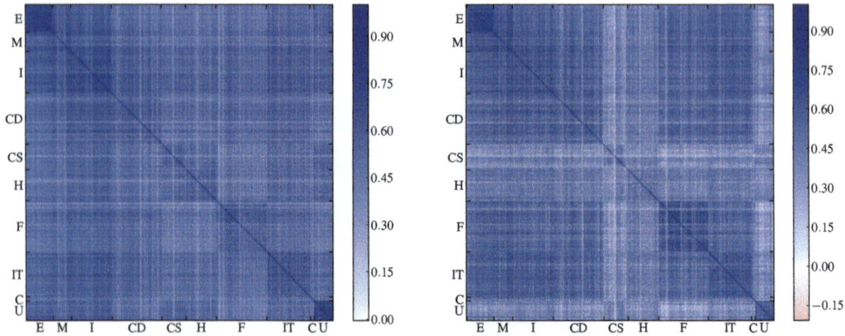

(a) Crisis (2008/10/15 - 2009/4/1, excluding stable period)

(b) Stable period (2009/1/1 - 2009/1/21)

Figure 2.7: Correlation matrix of the 2008–2009 crisis and the stable period during the crisis.

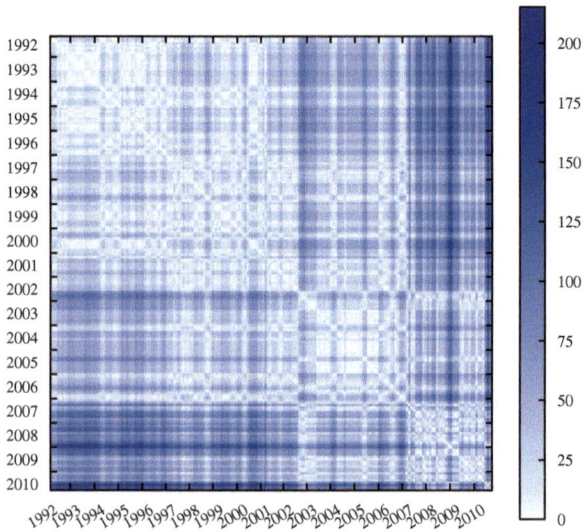

Figure 2.8: Similarity measure based on the difference of the correlation matrices' largest eigenvalues.

2.1.2.1 Alternative Measure: Difference of the Largest Eigenvalue of Correlation Matrices

A similar result can be achieved using a different approach. The largest eigenvalue λ_1 of the correlation matrix $\mathbf{C}$ describes the collective motion of all stocks. We can also define the similarity measure by the distance of these eigenvalues,

$$\zeta_{\text{alt}}^{(L)}(t_1, t_2) \equiv \left| \lambda_1(\mathbf{C}^{(L)}(t_1)) - \lambda_1(\mathbf{C}^{(L)}(t_2)) \right| \ . \tag{2.3}$$

Fig. 2.8 illustrates that this leads to an almost identical result. The advantage of this technique is that the noise in the correlation matrix only contributes to small eigenvalues (See, e.g., Refs. [10, 11]). Thus, by only taking into account the largest one, we can filter out a portion of the noise. However, this approach also presumes that the corresponding eigenvector does not change. Our results indicate that the largest eigenvalue almost remains constant, but this might not always be the case, especially in financial crises.

2.1.3 Application to Portfolio Optimization

The classification of market states, as carried out in the previous section, is only one of many possible applications of the similarity measure. As a practical example, we utilize the similarity measure as an instrument in risk management for the adaptive estimation of correlation matrices.

In this section, we first give a brief introduction to portfolio optimization. Then, we develop an estimator for calculating similarity-weighted correlation matrices. Eventually, we demonstrate the technique in an empirical study.

2.1.3.1 Modern Portfolio Theory

The *Modern Portfolio Theory (MPT)* or *Mean-Variance Portfolio optimization (MVO)*, developed by Markowitz in 1952 [52–54] is a standard approach in economics. It aims at reducing the overall risk of an portfolio Ω^2, as introduced in section 1.2.2, by calculating the fractions of wealth.

The assumption is that one can extrapolate the covariance matrix $\mathbf{\Sigma}$ based on historical data to a certain period in the future. This assumption does not only imply that the statistical dependence does not change, but also that the variances remain the same, i.e., they are assumed as constant. Moreover, as the variance, or the volatility is used as a risk measure, returns are considered as normal distributed returns. Heavy tails of the return distribution are neglected. Approaches with more suitable distribution functions have been made [55, 56] and higher statistical moments have been considered [57, 58].

Nevertheless, in today's financial world, the MPT is still widely used as an efficient and powerful tool in many areas mainly because of its simplicity. The MPT optimization takes three inputs:

1. The covariance matrix $\mathbf{\Sigma}$
2. The drift or *expected return* $\vec{\mu}$
3. The *desired portfolio return* R

The correlation matrix directly affects the output of the MPT. A recent study indicates that the risk of an optimized portfolio is lowest if the covariance matrix for the optimization is estimated accurately [59]. Thus, MTP represents an excellent technique to test new correlation estimation methods such as the similarity measure.

For the following considerations, let us assume a dimensionless, discrete time with time step $\Delta t = 1$. The value $V_p(t)$ of a portfolio of K assets at time t is defined as the linear combination of their prices $S^{(k)}(t)$ and corresponding fractions of wealth $w_k(t)$ at time t.

$$V_p(t) = \sum_{k=1}^{K} w_k(t) S^{(k)}(t) = \vec{w}^{\dagger}(t) \vec{S}(t) \ , \tag{2.4}$$

where $\vec{S}(t)$ refers to the vector of prices $S^{(k)}(t)$ with K components and $\vec{w}(t)$ contains the fractions of wealth. We include the argument t to the factions of wealth to emphasize that they change in time. The fractions of wealth $w_k(t)$ are normalized and dimensionless.

As discussed in the introduction, given the covariance matrix $\mathbf{\Sigma}^{(T)}(t)$ estimated in the interval $[T-t, t]$ and the fractions of wealth, the portfolio risk is defined as

$$\Omega^2(t) = \sum_{k,l=1}^{K} \Sigma_{kl}^{(T)}(t) w_k(t) w_l(t) = \vec{w}^{\dagger}(t) \mathbf{\Sigma}^{(T)}(t) \vec{w}(t) \ . \tag{2.5}$$

In MTP, the covariance matrix is often separated into the correlation matrix $\mathbf{C}$ and the standard deviations $\vec{\sigma}$. This is due to the fact that the variance changes rapidly, whereas the correlation must be estimated on a long time horizon. Let us estimate the correlation matrices on time horizon T and the standard deviations on time horizon T_σ. with $\mathbf{S}^{(T_\sigma)}(t) = \mathrm{diag}(\vec{\sigma}^{(T_\sigma)}(t))$, we write

$$\mathbf{\Sigma}^{(T,T_\sigma)}(t) = \mathbf{S}^{(T_\sigma)}(t)\mathbf{C}^{(T)}(t)\mathbf{S}^{(T_\sigma)}(t) \ , \tag{2.6}$$

where $\mathbf{C}^{(T)}(t)$ refers to the correlation matrix estimated in the interval $[T-t,t]$. Hence, the portfolio risk can also be written as

$$\Omega^2(t) = \sum_{k,l=1}^{K} C_{kl}^{(T)}(t)\sigma_k^{(T_\sigma)}(t)\sigma_l^{(T_\sigma)}(t)w_k(t)w_l(t) \ . \tag{2.7}$$

For the sake of simplicity, in the following we assume $T = T_\sigma$ and thus denote covariance matrix with $\mathbf{\Sigma}^{(T)}(t)$.

The second input parameter of MTP, the drift $\vec{\mu}$, can be estimated, e.g., on historical data. Let us consider that we estimate the drift based on the portfolio's components historical returns on the time interval $[t-T_\mu,t]$. However, we omit the index T_μ for a simpler notation.

MPT aims for the minimization of $\Omega^2(t)$ under certain constraints, for example $w_k(t) \geq 0$, if short selling is prohibited. Short selling is a concept in economics in which a portfolio manager can sell assets although he is not in their possession. This corresponds to negative fractions of wealth. Although being controversially discussed, short selling represents a powerful tool in risk management and is very common to encounter.

In our case, we assume short selling to be allowed. Then we only have one constraint, the *budget constraint*. This means that the total amount of capital, which is spent on the securities is fixed. In other words, the sum over all fractions of wealth needs to be a constant. If the fractions of wealth are normalized, this constant becomes one,

$$\sum_{k=1}^{K} w_k(t) = \vec{w}^\dagger(t)\mathbb{1} = 1 \ , \tag{2.8}$$

where $\mathbb{1}$ is a vector with unity in all entries. The reduction of the portfolio risk then leads to the optimization problem

$$\min_{\vec{w}(t)} \left\{ \frac{1}{2}\vec{w}^\dagger(t)\mathbf{\Sigma}^{(T)}(t)\vec{w}(t) - \gamma\vec{w}^\dagger(t)\vec{\mu}(t) \;\middle|\; \vec{w}^\dagger(t)\mathbb{1} = 1 \right\} \ . \tag{2.9}$$

Here, γ is called the *risk tolerance parameter.* It gives a measure of the willingness of the investor to face more risk in order to achieve higher portfolio returns. We will show later that γ can be mapped to the desired portfolio return R. The optimization problem can be solved using Lagrange multipliers λ and γ for the constraints. Thus, we have the Lagrangian

$$L = \frac{1}{2}\vec{w}^\dagger(t)\mathbf{\Sigma}^{(T)}(t)\vec{w}(t) - \gamma\vec{w}^\dagger(t)\vec{\mu}(t) + \lambda(\vec{w}(t)\dagger\mathbb{1} - 1) \tag{2.10}$$

with the gradient on the factions of wealth,

$$\frac{\partial}{\partial\vec{w}(t)}L = \mathbf{\Sigma}^{(T)}(t)\vec{w}(t) - \gamma\vec{\mu}(t) + \lambda\mathbb{1} \; . \tag{2.11}$$

By $\partial/\partial\vec{w}(t)L = 0$ we obtain optimal fractions of wealth $\vec{w}_{\text{opt}}(t)$,

$$0 = \mathbf{\Sigma}^{(T)}(t)\vec{w}_{\text{opt}}(t) - \gamma\vec{\mu}(t) + \lambda\mathbb{1} \tag{2.12}$$

$$\vec{w}_{\text{opt}}(t) = \gamma(\mathbf{\Sigma}^{(T)}(t))^{-1}\vec{\mu}(t) - \lambda(\mathbf{\Sigma}^{(T)}(t))^{-1}\mathbb{1} \; . \tag{2.13}$$

Moreover, the budget constraint $\mathbb{1}^\dagger\vec{w}(t) = 1$ gives

$$1 = \gamma\mathbb{1}^\dagger(\mathbf{\Sigma}^{(T)}(t))^{-1}\vec{\mu}(t) - \lambda\mathbb{1}^\dagger(\mathbf{\Sigma}^{(T)}(t))^{-1}\mathbb{1} \tag{2.14}$$

$$\lambda = \frac{\gamma\mathbb{1}^\dagger(\mathbf{\Sigma}^{(T)}(t))^{-1}\vec{\mu}(t) - 1}{\mathbb{1}^\dagger(\mathbf{\Sigma}^{(T)}(t))^{-1}\mathbb{1}} \; . \tag{2.15}$$

Inserting Eq. (2.15) into Eq. (2.13) leads to

$$\vec{w}_{\text{opt}}(t) = \gamma(\mathbf{\Sigma}^{(T)}(t))^{-1}\vec{\mu}(t) - \frac{\gamma\mathbb{1}^\dagger(\mathbf{\Sigma}^{(T)}(t))^{-1}\vec{\mu}(t) - 1}{\mathbb{1}^\dagger(\mathbf{\Sigma}^{(T)}(t))^{-1}\mathbb{1}}(\mathbf{\Sigma}^{(T)}(t))^{-1}\mathbb{1} \tag{2.16}$$

$$= \gamma(\mathbf{\Sigma}^{(T)}(t))^{-1}\vec{\mu}(t) - \gamma(\mathbf{\Sigma}^{(T)}(t))^{-1}\mathbb{1}\frac{\mathbb{1}^\dagger(\mathbf{\Sigma}^{(T)}(t))^{-1}\vec{\mu}(t)}{\mathbb{1}^\dagger(\mathbf{\Sigma}^{(T)}(t))^{-1}\mathbb{1}} + \frac{(\mathbf{\Sigma}^{(T)}(t))^{-1}\mathbb{1}}{\mathbb{1}^\dagger(\mathbf{\Sigma}^{(T)}(t))^{-1}\mathbb{1}} \tag{2.17}$$

$$= \gamma(\mathbf{\Sigma}^{(T)}(t))^{-1}\left(\vec{\mu}(t) - \mathbb{1}\frac{\mathbb{1}^\dagger(\mathbf{\Sigma}^{(T)}(t))^{-1}\vec{\mu}(t)}{\mathbb{1}^\dagger(\mathbf{\Sigma}^{(T)}(t))^{-1}\mathbb{1}}\right) + \frac{(\mathbf{\Sigma}^{(T)}(t))^{-1}\mathbb{1}}{\mathbb{1}^\dagger(\mathbf{\Sigma}^{(T)}(t))^{-1}\mathbb{1}} \; . \tag{2.18}$$

Using $\alpha = \mathbb{1}^\dagger(\mathbf{\Sigma}^{(T)}(t))^{-1}\vec{\mu}(t)$ and $\beta = \mathbb{1}^\dagger(\mathbf{\Sigma}^{(T)}(t))^{-1}\mathbb{1}$, this expression can be simplified to

$$\vec{w}_{\text{opt}}(t) = \frac{(\mathbf{\Sigma}^{(T)}(t))^{-1}\mathbb{1}}{\beta} + \gamma(\mathbf{\Sigma}^{(T)}(t))^{-1}\left(\vec{\mu}(t) - \frac{\mathbb{1}\alpha}{\beta}\right) \; . \tag{2.19}$$

However, in this expression, the willingness of the investor to face more risk, i.e., aim at higher returns is controlled by γ. We can map γ to the desired portfolio return R, by expressing it through the optimal fractions of wealth and the expected individual returns, the drift $\vec{\mu}(t)$. We can express the desired portfolio return as

$$R = \vec{w}_{\mathrm{opt}}^{\dagger}(t)\vec{\mu}(t) = \vec{\mu}^{\dagger}(t)\vec{w}_{\mathrm{opt}}(t) \ . \tag{2.20}$$

Thus, by multiplying Eq. (2.19) with $\vec{\mu}^{\dagger}(t)$, we obtain

$$R = \frac{\alpha}{\beta} + \gamma \left(\vec{\mu}^{\dagger}(t) \left(\mathbf{\Sigma}^{(T)}(t) \right)^{-1} \vec{\mu}(t) - \frac{\alpha^2}{\beta} \right) \tag{2.21}$$

$$\gamma = \frac{R - \frac{\alpha}{\beta}}{\vec{\mu}^{\dagger}(t)(\mathbf{\Sigma}^{(T)}(t))^{-1}\vec{\mu}(t) - \frac{\alpha^2}{\beta}} \ . \tag{2.22}$$

Therefore, given a certain desired return, an estimation of the covariance matrix and the drift, the optimal portfolio weights can be calculated. A value of $\gamma = 0$ denotes no risk tolerance. In this case, the investor's only aim is to minimize the portfolio variance. The resulting portfolio is referred to as the *minimum variance portfolio (MVP)*. Large values of γ denote risk neutrality, i.e., the investor maximizes the desired portfolio return only. The latter represents the *target return portfolio (TRP)*.

The volatilities can be estimated on short historical time horizons or using autoregressive models (GARCH) or exponential weighted moving averages (EWMA). For details see, e.g., Refs. [60–62]. Further recent studies demonstrate how to predict the volatility based on higher order multi-scale statistics given by a hierarchical process [63, 64]. An overview on various volatility forecasting techniques is given in Ref. [65].

Generally, the quality of the estimated matrices increases with the length of the time series, i.e., the amount of data used. For small datasets the matrices have a large variance and may even be singular or indefinite. In financial context, however, using long time series results in biased estimates of the correlation structure, since the dependence of asset returns is not constant in time.

Good estimates of the correlation structure are the key in MVO. The problem is that standard estimators equally weight all parts of the dataset. By consequence, out-of-date and improper information highly affect the estimates. Here, we approach this problem by utilizing the similarity measure for the estimation of weighted covariance matrices. This estimator makes use of enough data to adequately limit its variance but – in order to minimize its bias – focuses only on parts of the data where the market is in similar market conditions.

To reduce the effects of time changing structures, common approaches in literature choose time intervals where the structures are approximately constant. Examples of such approaches are exponentially weighted estimators like the RiskMetrics estimators [66, 67]. Since these estimators only use a small part of the data, they show a large variance. Moreover, whenever the number of effectively used observations is not large compared to the number of time series, estimated correlation and covariance matrices may be regarded as dominated by randomness. As discussed in the introduction, a study indicates that 94% of the spectrum of estimated correlation matrices is equal to the spectrum of random matrices [11]. Only their largest eigenvalues may be estimated adequately.

Solutions to this issue involve reducing the dimensionality of the problems by imposing some structure on the correlations, e.g., by using factor models or shrinkage estimators as in [68] or by noise reduction techniques, e.g., Random Matrix Filtering [10, 11] or Power Mapping [26, 27]. Other approaches reduce the dimensionality by using conditional models for the correlation matrices [69].

With the availability of intraday high frequency financial data, it was expected that finer sampled data would effectively enlarge the datasets and improve estimates of parameters. However, when return data are observed on shorter time intervals, it is contaminated by market microstructure effects [70]. These effects influence the input parameters of MPT [71] and induce the Epps effect, discussed in chapter 3.

Since the amount of data for the estimation may only be increased by either considering a longer time period or by sampling on higher frequencies, the mentioned properties of financial time series limit the amount of usable data. Longer time intervals bias the estimators due to the non-stationarity of the market. Higher frequencies intensify the effects of the market microstructure on the estimators.

However, we can circumvent these limits using the similarity measure. We can enlarge the amount of usable data by adaptively including different parts of the time series with similar correlation structures into the estimator. The similarity measure enables us to construct a weighting scheme for correlation or covariance estimators that attaches high weights on similar parts of the data and suppresses distortions.

2.1.3.2 Calculating Weighted Correlation Matrices

The similarity measure $\zeta^{(L)}$ can serve as a weighting scheme for estimators of correlation or covariance matrices. With respect to the reference point t_0 the scheme inscribes high weights to periods where the market behaved in a similar manner. On the other hand, the periods in which the market behaved very differently are suppressed. Therefore, consider the adapted similarity measure

$$\tilde{\zeta}^{(L)}(t,t_0) = \frac{K-1}{K} - \zeta^{(L)}(t,t_0), \quad t \in [t_0 - T, t_0], \tag{2.23}$$

where T is the total number of considered time steps, i.e., the length of the time series. It is easily checked that $(K-1)/K$ represents the theoretical maximum possible value of $\zeta^{(L)}$. In case of identical market situations, we have $\tilde{\zeta}^{(L)} = 1$. The highest possible dissimilarity yields $\tilde{\zeta}^{(L)} = 0$.

We note that the matrices $\mathbf{C}^{(L)}$ in Eq. (2.1) are estimated with window length L on daily returns. Therefore, within the timespan $[t_0 - L, t_0]$, they share identical values with the matrix at $t = t_0$. $\tilde{\zeta}^{(L)}(t,t_0)$ is then dominated by the amount of identical values and not by the estimated similarity. Therefore, the similarity measure is not reliable within this region and is set to the maximum value of the other timespans, resulting in a corrected measure

$$\tilde{\zeta}^{*(L)}(t,t_0) = \begin{cases} \max(\tilde{\zeta}^{(L)}(t < t_0 - L, t_0)) & t \in [t_0 - L, t_0] \\ \tilde{\zeta}^{(L)}(t,t_0) & t \in [t_0 - T, t_0 - L[\; . \end{cases} \tag{2.24}$$

A properly normalized weighting scheme for the estimation of the correlation or covariance matrix, $\mathbf{C}^{(T)}(t_0)$ or $\mathbf{\Sigma}^{(T)}(t_0)$ at time $t = t_0$ is then given by

$$g(t,t_0,L) = \tilde{\zeta}^{*(L)}(t,t_0) \Bigg/ \left(\sum_{t'=t_0-T}^{t_0} \tilde{\zeta}^{*(L)}(t',t_0) \right) , \tag{2.25}$$

resulting in the weighted estimators

$$\widehat{\mathbf{C}}^{(T)}(t_0) = \sum_{t=t_0-T}^{t_0} g(t,t_0,L)\, \mathbf{C}^{(L)}(t) \quad \text{and} \tag{2.26}$$

$$\widehat{\mathbf{\Sigma}}^{(T)}(t_0) = \sum_{t=t_0-T}^{t_0} g(t,t_0,L)\, \mathbf{\Sigma}^{(L)}(t) \; . \tag{2.27}$$

$\mathbf{C}^{(L)}(t)$ and $\mathbf{\Sigma}^{(L)}(t)$ denote the correlation and covariance matrix of the interval $[t-L,t]$. For large T and time series with dynamic correlation structure, the weighting scheme should be restricted only to the s largest values of g. This leads to a complete suppression of dissimilar parts of the data. To accomplish this, let us denote the s-th largest value of g with q_s. The restricted scheme g_s is then given by

$$g_s(t,t_0,L) = |g-q_s|_+ \Big/ \sum_{t'=t_0-T}^{t} |g-q_s|_+ \ , \tag{2.28}$$

with

$$|g-q_s|_+ \begin{cases} g(t',t_0,L) & g(t',t_0,L) > q_s \\ 0 & \text{else.} \end{cases} \tag{2.29}$$

Using this estimator, we are able to calculate the correlation $\widehat{\mathbf{C}}^{(T)}(t_0)$ matrix, based on T matrices that have the highest similarity with the current situation, i.e., similar to $\mathbf{C}^{(L)}(t_0)$.

2.1.3.3 Empirical Study

Now we apply our estimator to financial data in the context of MPT optimization. We use the same dataset (i) of section 2.1.1, i.e., permanent constituents of the S&P 500 index that are included in the index from 1994 to the mid of 2010. From this dataset, we randomly choose 10 portfolio constellations of 100 stocks each, as listed in appendix A.6.

On every 14-th day in the period we compute the optimal fractions of wealth for the constellations regarding the two strategies TRP and MVP. The required covariance estimates are based on the restricted similarity-weighted estimator defined in Eqs. (2.27) and (2.28) and alternatively on the unweighted estimator. The matrices to calculate the similarity measure are based on moving windows of $L = 32$ trading days. The weighting scheme of the estimator includes the $s = 300$ most similar past days of the full dataset. In other words, we chose the 300 most similar correlation matrices from the year 1994 to the point of calculation. The unweighted estimator is based on a moving window of 300 days, i.e., T=300.

The vectors $\vec{\mu}$ and $\vec{\sigma}$ is estimated by the returns of the portfolio's stocks for every trading day from a moving window of $T_\mu = 14$ trading days. In the TRP case, the desired return R is adaptively chosen to be 5 percentage points above the average component of $\vec{\mu}$. The basic idea of this study is to calculate optimal portfolios for every 14 days of our dataset and to evaluate them over some investment horizon T' with respect to risk and realized return.

While the covariance matrix and the drift are estimated on data which is in the past in the point of optimization t_0, let us consider an investment period from day $t' = t_0 + 1$ to day $t = T'$ which is in the future from each point of optimization. This time window acts as an evaluation period. It enables us to evaluate how this portfolio would have been performed if it was optimized at $t = 0$. In this evaluation window the realized portfolio return with respect to the point of optimization t_0 is given by

$$R_p(t) = \frac{V_p(t) - V_p(t_0)}{V_p(t_0)} \quad , \quad t \in \{t_0 + 1, t_0 + 2, \ldots, T'\}. \tag{2.30}$$

Thereby we can simulate the portfolio performance as if an investor had performed the optimization at t_0. Now we define the *realized risk*, or realized variance RV of the portfolio as the variance of the realized portfolio return in the evaluation window,

$$RV = \mathrm{var}\left(R_p(t)\right) \quad , \quad t \in \{t_0 + 1, t_0 + 2, \ldots, T'\}. \tag{2.31}$$

This enables us to evaluate the optimization, which is an indication of how well the covariance matrix was estimated.

A recent study argues that minimum-variance portfolios outperform various other strategies of portfolio optimization, even with respect to their return [72]. By contrast, Ref. [73] raises the question whether portfolio optimization pays out at all. In their results, optimized portfolios do not significantly outperform naively diversified portfolios, i.e., portfolios where the same amount $1/K$ is invested in K assets. We therefore include this naive portfolio in our study, even though the naive portfolio does not depend on estimators of correlation or covariance. The portfolio strategies MVP and TRP allow ranking the estimators of the covariance structure according to the portfolio performance, whereas the outcomes of the naive portfolio confirm the overall plausibility of the results.

The evaluation results of realized risk and returns are shown in Figs. 2.9 and 2.10. The evaluation periods are 14, 28 and 56 trading days in order to analyze the stability of the obtained portfolios. The results shown are averages of the 10 portfolio constellations. The figures illustrate that the naive portfolio performs worst in terms of risk, especially during the financial

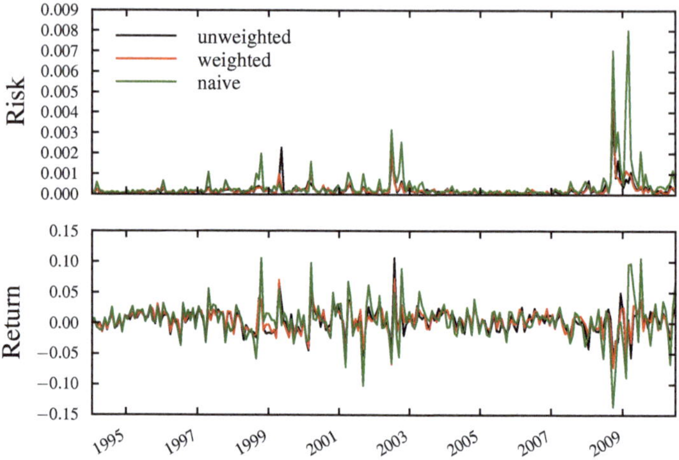

(a) 14 day evaluation

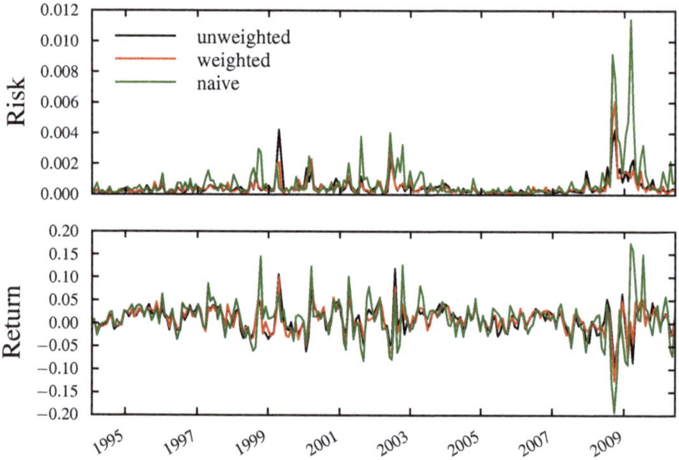

(b) 28 day evaluation

Figure 2.9: Average realized return and realized risk in mean-variance portfolio optimization for a target return of 5% above the market drift and different evaluation windows. The results are compared to a naive portfolio as a reference.

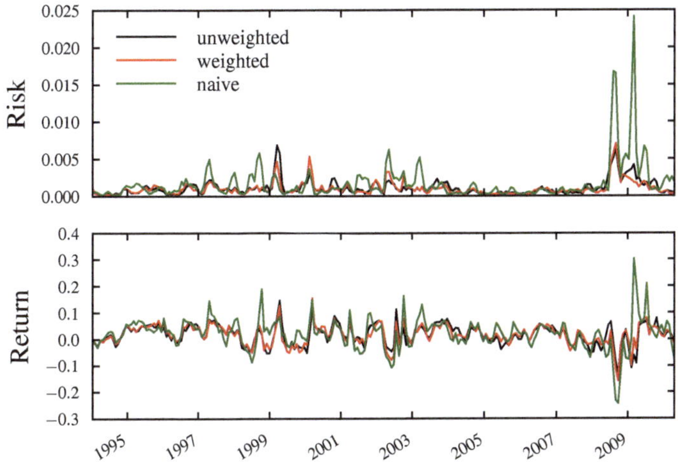

(c) 56 day evaluation

Figure 2.9: (continued)

		Optimization type		
Evaluation		unweighted	weighted	naive
14 day	Risk	0.00022	0.00021	0.00041
	Return	0.00398	0.00381	0.00472
28 day	Risk	0.00048	0.00046	0.00081
	Return	0.00802	0.00820	0.00999
56 day	Risk	0.00107	0.00099	0.00170
	Return	0.01632	0.01699	0.02127

Table 2.2: Average realized return and realized risk in mean-variance portfolio optimization for a target return of 5% above the market drift and different evaluation windows. The results are compared to a naive portfolio as a reference.

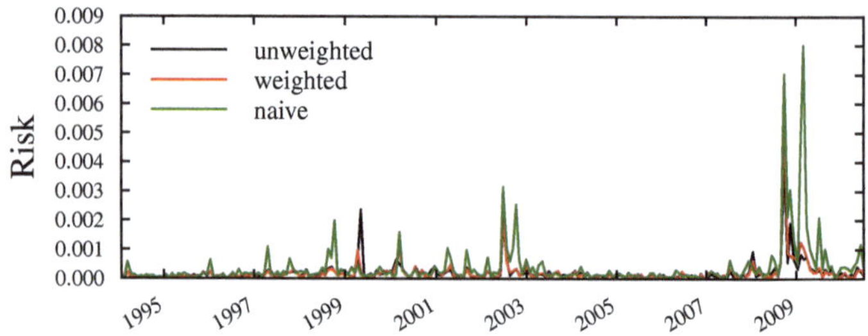

(a) 14 day evaluation

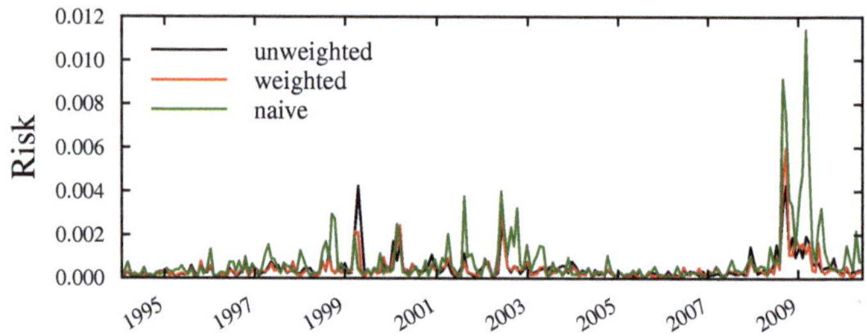

(b) 28 day evaluation

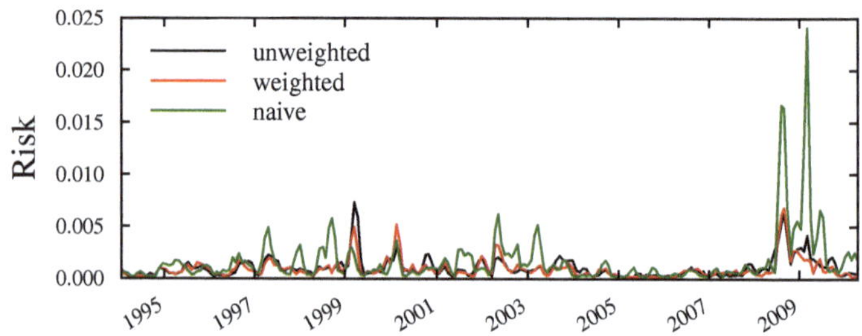

(c) 56 day evaluation

Figure 2.10: Average realized risk in mean-variance portfolio optimization for the minimum variance portfolio and different evaluation windows. The last column provides a comparison to the naive portfolio.

	Optimization type		
Evaluation	unweighted	weighted	naive
14 day	0.00022	0.00021	0.00041
28 day	0.00047	0.00045	0.00081
56 day	0.00106	0.00098	0.00170

Table 2.3: Average realized risk in mean-variance portfolio optimization for the minimum variance portfolio and different evaluation windows. The last column provides a comparison to the naive portfolio.

crisis between 2008-2009. At that time, the incorporation of the correlation structure into the portfolio weights pays out. Realized risk of the optimized portfolios consistently lies below the realized risk of the naive portfolios whereas the similarity-weighted scheme obtains the best results. The results are robust for the considered investment horizons, which is shown in Tabs. 2.2 and 2.3 in more detail.

In both cases, in the minimum variance portfolio (MVP) as well as in the 5% above market drift portfolio (TRP) perform about equally well when evaluating on a 14 day window. However, when prolonging the evaluation window and thereby testing for portfolio stability, the similarity-weighting significantly reduces the realized risk. Moreover, the TRP case reveals that the realized return could be improved compared to the unweighted optimization, although the naive portfolio features an even higher return. The outperforming by the naive portfolio in terms of realized returns is a common phenomenon. The reason is that it is very difficult to estimate the drift vector $\vec{\mu}$ properly. The non-stationarity features of the market make the drift estimation one of the most challenging tasks in praxis. Because of the short time horizon, the similarity measure is not applicable here. A better estimation of $\vec{\mu}$ would result in a realized return that is closer to the desired return. However, the main motivation behind MVO is the minimization of the portfolio risk, not the portfolio return. As we estimate $\vec{\mu}$ in the same way for both, the weighted and unweighted portfolio, we can compare the results or in other words, the impact of the estimation of $\vec{\mu}$ is identical for both portfolios.

2.1.4 Summary

We introduced a measure to quantify structural similarities of a market's correlation structure at two different points in time. In an empirical study, we demonstrated the power of this measure that discloses a general dynamics of a financial market. By providing a simple instrument to identify similarities to previous states during an upcoming crisis, one can judge the current situation properly and be prepared to react if the crisis materializes. Another indication for a crisis is given when the correlation structure undergoes rapid changes.

Using the similarity measure we were able to classify several typical market states between which the market jumps back and forth. Some of these states can easily be identified in the similarity measure. However, there are several states in which the market only stays for a short period. Thus, these states are sparsely embedded in time. With a clustering analysis, we were able to identify these states and disclose a detailed dynamics of the market's state.

A further possible application of the similarity measure is portfolio optimization. Given the similarity measure, the portfolio manager is aware of periods in which the market behaved completely differently and thus can choose not to include them in his calculations. He/She can furthermore identify regions in which the market behaved similarly and refer to these regions when estimating the correlation matrix. In an empirical study, we demonstrated that the utilization of the similarity measure in portfolio optimization leads to a significant reduction of risk.

2.2 A Copula Approach to Statistical Dependence of Stock Returns

The measurement of statistical dependence is often broken down to the calculation of correlations, such as the Pearson coefficient or the Spearman coefficient [74]. Correlation coefficients are widely used in various disciplines of science. They are very common in financial modeling, e.g., in the Capital Assets Pricing Model (CAPM) [75], Noh's model [76] (see chapter 3) or portfolio optimization, as discussed in section 2.1.3.

The Pearson correlation coefficient, however, only accounts for linear statistical dependence assuming that the observables are nearly normal distributed. Due to the central limit theorem, this might be justified in some cases, but often the statistical dependence is much more complex. In these cases, the statistical dependence cannot be represented by a single number such as a correlation coefficient. This is, e.g, the case if the statistical dependence grows with the absolute value of an observable.

The joint probability distribution of the observables holds all information of the statistical dependence. For example, the cumulative joint probability distribution of two random variables X and Y reads

$$F_{X,Y}(x,y) = P(X \leq x, Y \leq y). \tag{2.32}$$

It gives the probability P, that $X \leq x$ while $Y \leq y$. The joint probability density function is given by

$$f_{X,Y}(x,y) = \frac{\partial^2}{\partial x \partial y} F_{X,Y}(x,y). \tag{2.33}$$

However, a joint probability distribution also contains the individual marginal probability distributions. These can have different shapes depending on the underlying process. The statistical dependence of different systems usually cannot be directly compared with this approach.

Copulae, first introduced by Sklar in 1959 [77, 78], permit a separation between the pure statistical dependence and the marginal probability distributions. This allows comparing the statistical dependence of diverse systems.

The use of copulae is well established in statistics and finance. There are many classes of analytical copula functions that meet various properties [79]. Several studies of financial markets are devoted to developing suitable copulae or fitting existing ones to empirical data [80–82] or are based on a small subset of assets [83]. In this study, we chose a different approach. We perform a large-scale empirical study to disclose the structure of the average pairwise copula of the US stock returns. As the copula does not depend on the shape of the return distribution, we are able to average over the various copulae of different stock pairs even though the shape of their corresponding marginal distributions may differ, i.e., exhibits stronger or weaker tails. In particular, we study the intraday stock market returns of the 428 continuous S&P 500 constituents in 2007–2010 based on intraday data from the New York Stock Exchange's TAQ database [43]. Over 12 billion single transactions are analyzed.

This section is organized as follows. After introducing the concept of copulae in section 2.2.1 we calculate the average pairwise copula in section 2.2.2. The dynamics of this copula is discussed in section 2.2.3. We summarize the results in section 2.2.4.

2.2.1 Copulae

The basic concept is simple: Let a and b be two random variables with probability densities $f_a(x)$ and $f_b(x)$ and cumulative distributions $F_a(x)$ and $F_b(x)$, with

$$\int_{-\infty}^{+\infty} f_a(x)dx = 1 \ , \tag{2.34}$$

$$F_a(x) = \int_{-\infty}^{x} f_a(x') \, dx' \ , \tag{2.35}$$

and analogously for b. Further, let $f_{a,b}(x,y)$ be the joint probability density and $F_{a,b}(x,y)$ be the joint cumulative distribution of a and b. The inverse cumulative distribution function F^{-1} is called the quantile function. For example, $F_a^{-1}(0.05)$ represents the value which 5% of all random samples are smaller or equal to. This evidently gives,

$$F_a\left(F_a^{-1}(\alpha)\right) = \alpha \ . \tag{2.36}$$

$F^{-1}(\alpha)$ is also called the α-quantile. The copula $\mathrm{Cop}_{a,b}(u,v)$ is defined as the cumulative joint distribution of quantiles,

$$\mathrm{Cop}_{a,b}(u,v) = F_{a,b}\left(F_a^{-1}(u), F_b^{-1}(v)\right) \ . \tag{2.37}$$

The copula density $\text{cop}_{a,b}(u,v)$ is consequently defined by

$$\text{cop}_{a,b}(u,v) = \frac{\partial^2}{\partial u \partial v}\text{Cop}_{a,b}(u,v) \ . \tag{2.38}$$

As the quantile functions F^{-1} are scale free, the copula does not depend on the underlying marginal distributions. It only contains the pure statistical dependence. Thus, by obtaining the appropriate copula of a system, one can simply interchange the marginal distributions without any changes in the copula. This is very useful if the marginal distributions change for some reason, but the statistical dependence remains the same. We can rebuild the joint cumulative distribution from the copula and the individual distributions by

$$F_{a,b}(x,y) = \text{Cop}_{a,b}\left(F_a(x), F_b(y)\right) . \tag{2.39}$$

2.2.2 Average Copula

To calculate the cumulative copula from empirical data of two return time series r_1 and r_2, we use

$$\text{Cop}_{1,2}(u,v) = \frac{1}{T}\sum_{t=1}^{T} 1_{\text{U}}(r^{(1)}(t))1_{\text{V}}(r^{(2)}(t)) \ , \tag{2.40}$$

where T is the length of the time series. 1_{U} and 1_{V} are indicator functions relating to the sets

$$U = \left\{x \mid x \leq F_1^{-1}(u)\right\} \ , \tag{2.41}$$

$$V = \left\{y \mid y \leq F_2^{-1}(v)\right\} \ . \tag{2.42}$$

Thus 1_U is given by

$$1_U(r_1(t)) = \begin{cases} 1 & r^{(1)}(t) \in U \\ 0 & r^{(1)}(t) \notin U \end{cases} \tag{2.43}$$

and analogously for 1_V. On empirical data, the aforementioned quantile function F^{-1} is given by

$$F_1^{-1}(u) = \begin{cases} \inf\left\{x \mid F_1(x) \geq u\right\} & 0 < u \leq 1 \\ \sup\left\{x \mid F_1(x) = u\right\} & u = 0 \end{cases} , \tag{2.44}$$

and analogously for r_2. We define $F_1(x)$ empirically as the percentage of the portion that is smaller or equal to x compared to the total amount of values. When calculating the empirical copula density, it is useful to first define a resolution of the 2D grid, e.g. $m = 50$. On this $m \times m$ grid, we can calculate the copula by

$$\mathrm{cop}_{1,2}\left(\frac{i}{m}, \frac{j}{m}\right) = \frac{1}{T}\sum_{t=1}^{T} 1_{\bar{\mathrm{U}}_\mathrm{i}}(r^{(1)}(t)) \times 1_{\bar{\mathrm{V}}_\mathrm{j}}(r^{(2)}(t)) \quad , \quad i, j \in 1 \ldots m \tag{2.45}$$

with

$$\bar{U}_i = \left\{ x \;\middle|\; F_1^{-1}\left(\frac{i-1}{m}\right) < x \leq F_1^{-1}\left(\frac{i}{m}\right) \right\} , \tag{2.46}$$

$$\bar{V}_j = \left\{ y \;\middle|\; F_2^{-1}\left(\frac{j-1}{m}\right) < y \leq F_2^{-1}\left(\frac{j}{m}\right) \right\} . \tag{2.47}$$

An accurate estimation of the copula density requires a large amount of data points. Thus, we estimate the average copula using intraday data. The analysis is performed for return intervals from 1 minute to 4 hours. Results are shown in Fig. 2.11

We obtain a similar copula for all return intervals. This is surprising because it is well-known that the shape of the marginal return distribution changes towards small return intervals – the tails of the distributions become stronger [20, 21]. However, apparently this does not change the statistical dependence in the same magnitude.

For very small return intervals, we find surprising phenomenon. The buckle in the center of Fig 2.11f is an artifact caused by the discretization of stock returns. In our calculations, we use a resolution of 0.02. On small return intervals, the number of certain returns can be larger than the amount of values per grid step. In our case, this is 2% of the total values in the marginal distributions. If the amount of returns on a certain value is higher, it also contributes to the following grid steps. All returns within this interval in one marginal distribution correspond to the same quantile in the other margin distribution, resulting in a peak of the copula. This effect mainly occurs in the center of the marginal distributions and if the exposure to the discretization is high, i.e., on small return intervals. As the position of the peaks differs from copula to copula, we see a smoothed peak in the average pairwise copula.

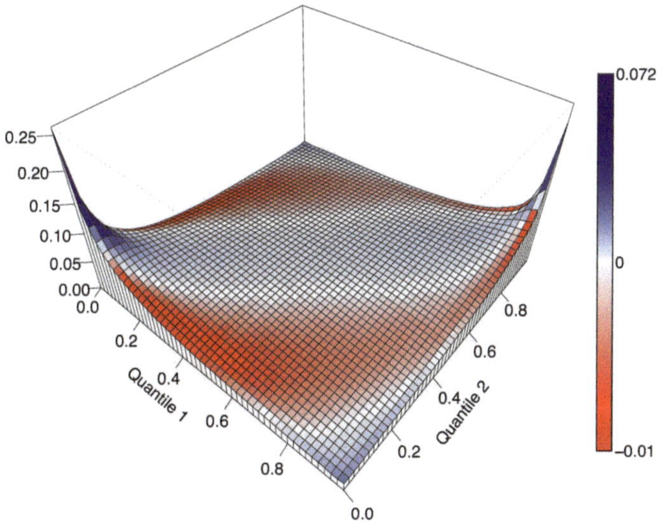

(a) $\Delta t = 240$ min

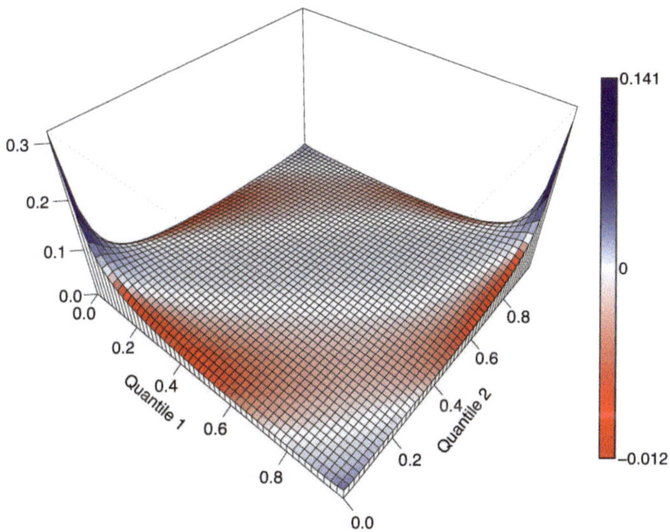

(b) $\Delta t = 60$ min

Figure 2.11: Average pairwise copula of the S&P 500 stock returns in 2007–2010 for various return intervals. The z-axis in permille. The color shading illustrates the difference to the Gaussian copula (positive values mean that Gaussian copula is less dense).

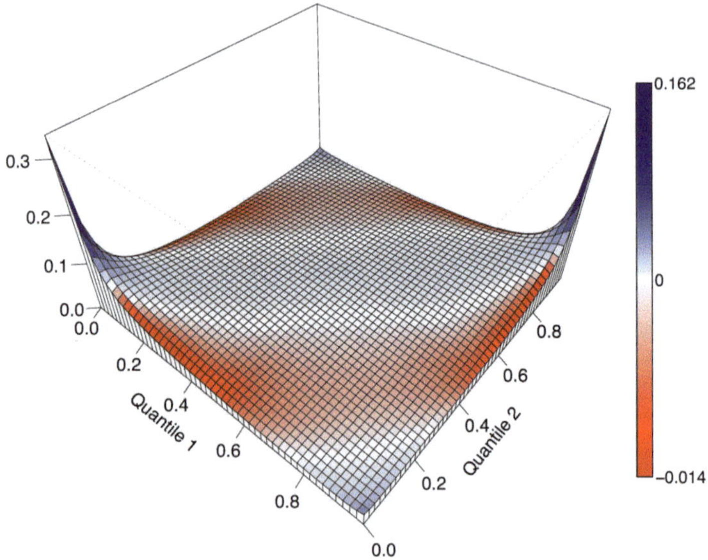

(c) $\Delta t = 30$min

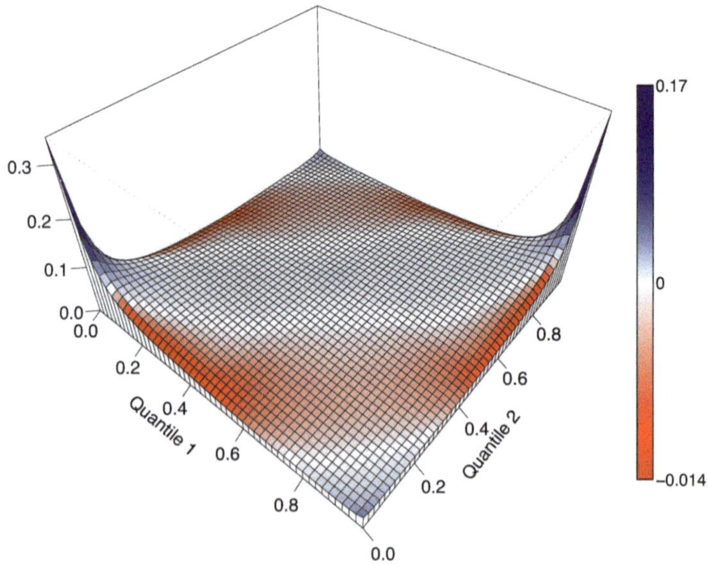

(d) $\Delta t = 20$ min

Figure 2.11: (continued)

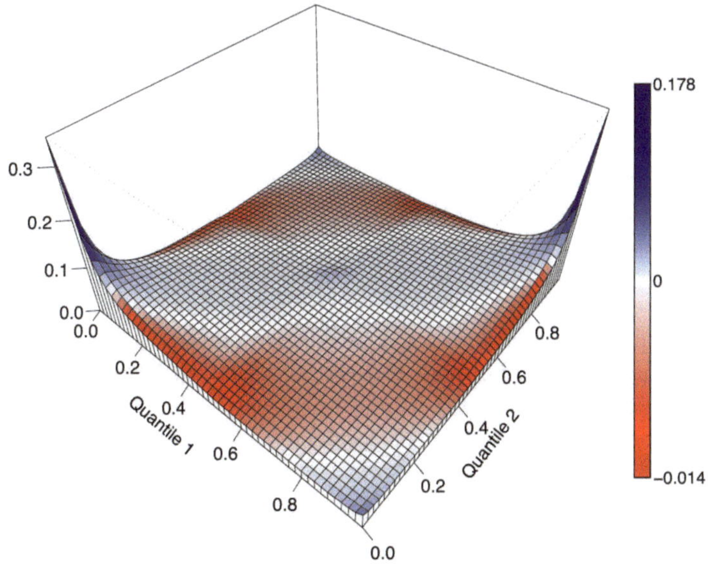

(e) $\Delta t = 10$ min, The discretization artifact emerges in the center.

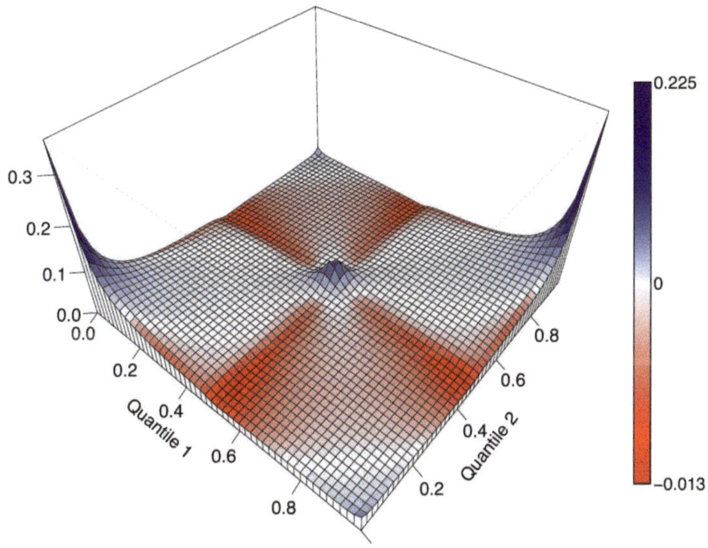

(f) $\Delta t = 1$ min, The discretization artifact is clearly visible

Figure 2.11: (continued)

The stability of the copula is especially remarkable because due to the Epps effect financial correlations decline towards smaller return intervals (see chapter 3). However, beside the buckle in the center the structure of the copula remains similar. The buckle corresponds to the amount of discretization. One can interpret it as the distortion of the statistical dependence due to discretization. In the copula, we can easily identify and isolate this effect. For a correlation coefficient which provides a single value instead of a 2D-function, this task becomes more complex. In section 3.2, we discuss how to compensate for the impact of discretization in a Pearson correlation coefficient.

The copulae have high density in the outer quantiles. This corresponds to a higher correlation in the tails of the return distribution than in it's center. This is often referred to as *tail dependence* [84–86]. Our results indicate that on average, the upper tail dependence is stronger than the lower tail dependence. For comparison, the average difference to the Gaussian copula (which is implied by most correlation coefficients) is illustrated in Fig. 2.11. The (standard normal) Gaussian Copula is given by

$$\mathrm{Cop}_c(u,v) = F_c(F^{-1}(u), F^{-1}(v)) \ , \tag{2.48}$$

$$\mathrm{cop}_c(u,v) = \frac{f_c(F^{-1}(u), F^{-1}(v))}{f(F^{-1}(u))f(F^{-1}(v))} \ . \tag{2.49}$$

Here, f_c and F_c refer to the bivariate standard normal probability density and cumulative distribution with correlation c. f is the univariate standard normal probability density, while F^{-1} is the corresponding quantile function. To calculate the average difference d, we have to calculate the Gaussian copula based on all coefficients of the correlation matrix $\mathbf{C}$, based on $K = 428$ stocks and subtract it from the empirical copula,

$$d(u,v) = \frac{\sum\limits_{i=1}^{K}\sum\limits_{j=i+1}^{K}\left(\mathrm{cop}_{i,j}(u,v) - \mathrm{cop}_{C_{i,j}}(u,v)\right)}{K(K-1)/2} \ . \tag{2.50}$$

This gives us information about how erroneous the dependence is estimated if implying a Gaussian copula. The empirical copula exhibits a stronger dependence than the Gaussian copula. The probability of correlated extreme events significantly is underestimated. Furthermore, on small return intervals the tail dependence is equally strong for both tails. Towards larger return intervals, the lower tail dependence becomes stronger than the upper tail dependence. This might be caused by the market reacting more

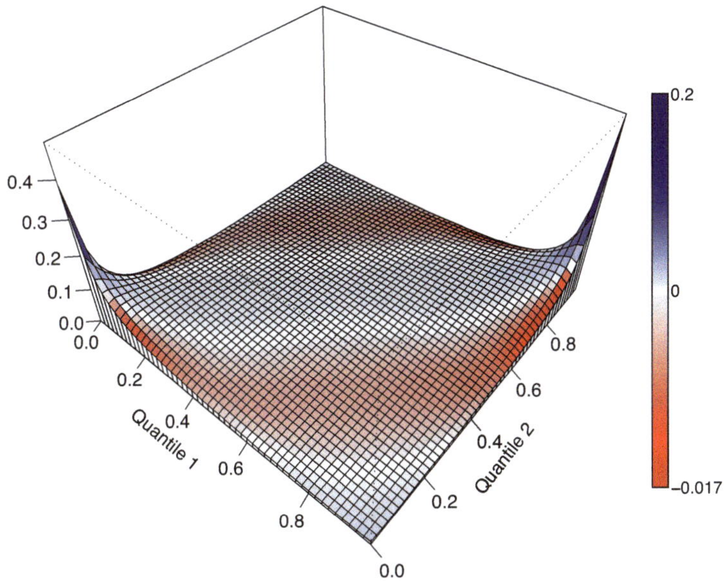

Figure 2.12: Average pairwise copula of 60min stock returns during the crisis period from 2008/10/15 to 2009/4/1. The color shading illustrates the difference to the Gaussian copula during this period.

severely on bad news than on good news [87]. We discuss this in more detail in the next section. Another feature of the empirical copula is the relatively high density in the (0,1) and (1,0) corners (except during the 2008–2009 crisis), indicating the presence of anti-correlated extreme events.

During the financial crisis period from Oct 2008 to Apr 2009, the difference to the Gaussian copula increases, as Fig 2.12 illustrates. The assumption of the Gaussian Copula would have been a dramatic mistake during this period. The Gaussian copula is even being discussed for having a main impact on the financial crisis [30]. In addition to an overall strong correlation level during the crisis, as discussed in section 2.1.2, a large portion of the statistical dependence lies in the tails of the marginal distributions. The probability of correlated extreme events is very high. Surprisingly the copula during the crisis exhibits a stronger positive tail dependence than negative tail dependence.

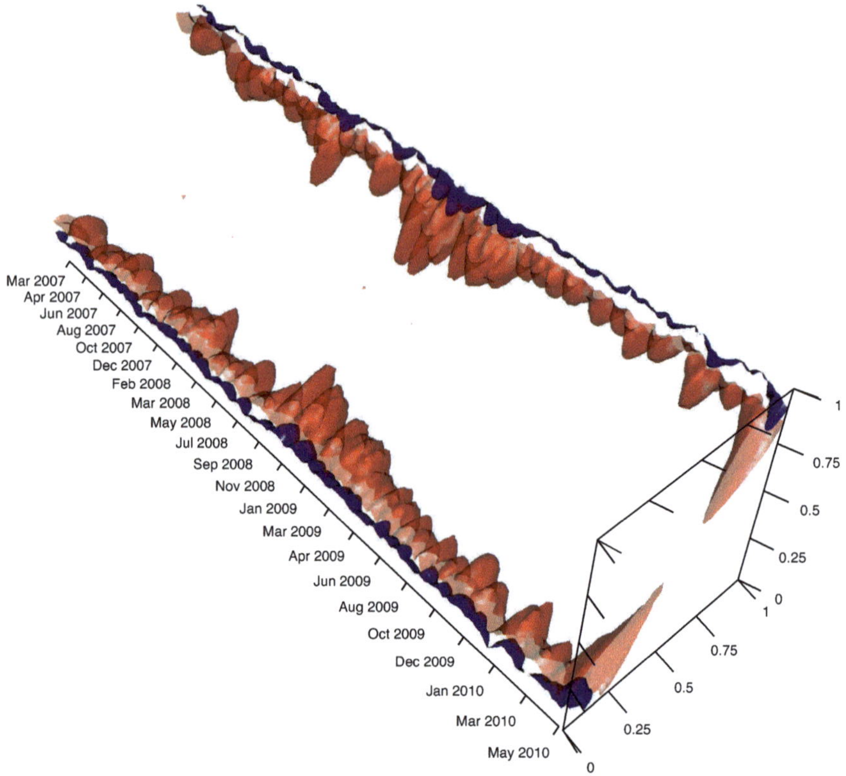

Figure 2.13: Evolution of the S&P 500 stocks' average pairwise copula density. The isosurfaces correspond to a probability of 0.1‰ (blue) and 0.05‰ (red). The density in the tails is very high.

2.2.3 Dynamics of the Copula

It is evident that statistical dependencies of financial assets change in time. For example, this can be caused by microeconomic influences, changing political factors or herding effects. We approached this matter with a similarity measure and the identification of market states in section 2.1. Earlier studies address this issue by discussing the dynamics of correlations [40–42, 88]. Here, we discuss a different approach on this topic. We perform a empirical study of the changes in the average pairwise copula. We calculate the average copula within 2-week periods within the 2007-2010 period based on 1-hour returns. Results are shown in Fig. 2.13. To illustrate the structural changes of the copula, we plot the isosurfaces in the tail regions. We discover that the tail dependence is stronger during financial crashes, such as from Oct 2008 to Feb 2010. However, the fluctuations of the tail dependence are very large. It reflects the current market's situation in a sensible manner.

Often financial crashes are accompanied by overall very large correlations. This raises the question if there is some dependence between the market's average correlation level and the tail dependence. To obtain an insight into this question we compare the average correlation coefficient of the whole market in each 2-week period to the tail dependence. As correlation coefficients are still widely used, this maps a correlation coefficient to one of the most important features of the copula.

To quantify this tail dependence, we calculate the probability of two returns to be simultaneous above or below a certain quantile α. This simple upper and lower tail dependence coefficient is given by

$$\lambda_l(\alpha) = \mathrm{Cop}(\alpha, \alpha) \ , \tag{2.51}$$

$$\lambda_u(\alpha) = 1 - \mathrm{Cop}(1-\alpha, 1-\alpha) \ . \tag{2.52}$$

More advanced tail dependence coefficients are, e.g., discussed in Ref. [86]. However, as we only examine the difference between the empirical copula and the Gaussian copula, we restrict ourselves to this measure. We perform the analysis for return intervals from 30 minutes to two hours. Results are shown in Fig. 2.14. We find a very strong relation of the tail dependence and the average correlation coefficient. For comparison we build the average tail dependence coefficients λ_l and λ_u of the Gaussian copula, given by

$$\lambda_l = \lambda_u = \mathrm{Cop}_c(\alpha, \alpha) \ . \tag{2.53}$$

To calculate the average Gaussian tail dependence, for each 2-week period, we calculate the tail dependence of the Gaussian copula based on the correlation matrix' entries $C_{i,j}$ of this period,

$$\left\langle \lambda_l^{(Gauss)} \right\rangle = \left\langle \lambda_u^{(Gauss)} \right\rangle = \frac{\sum\limits_{i=1}^{K} \sum\limits_{j=i+1}^{K} \left(\mathrm{Cop}_{C_{i,j}}(\alpha, \alpha) \right)}{K(K-1)/2} \ . \tag{2.54}$$

This gives the opportunity to compare how the tail dependence is overall misjudged, if using correlation coefficients or the Gaussian copula.

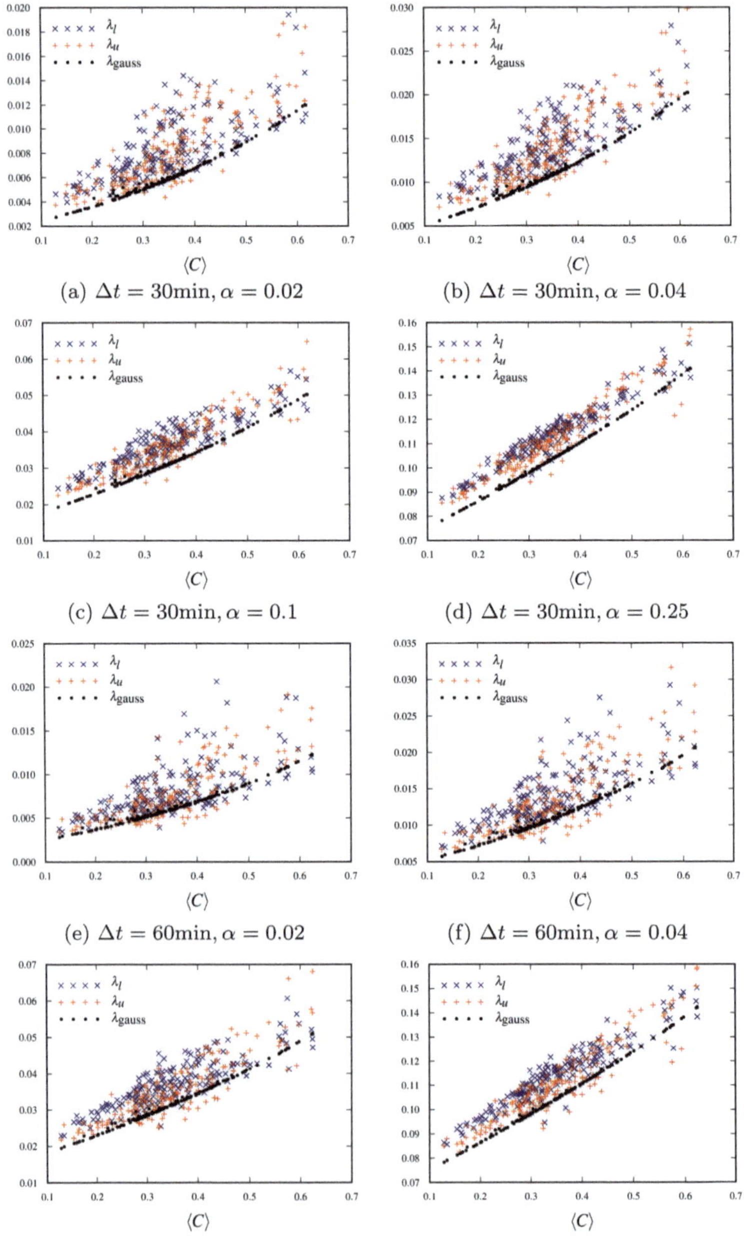

(a) $\Delta t = 30\text{min}, \alpha = 0.02$
(b) $\Delta t = 30\text{min}, \alpha = 0.04$
(c) $\Delta t = 30\text{min}, \alpha = 0.1$
(d) $\Delta t = 30\text{min}, \alpha = 0.25$
(e) $\Delta t = 60\text{min}, \alpha = 0.02$
(f) $\Delta t = 60\text{min}, \alpha = 0.04$
(g) $\Delta t = 60\text{min}, \alpha = 0.1$
(h) $\Delta t = 60\text{min}, \alpha = 0.25$

Figure 2.14: Relation between tail dependence and average correlation level for different quantiles α and return intervals Δt.

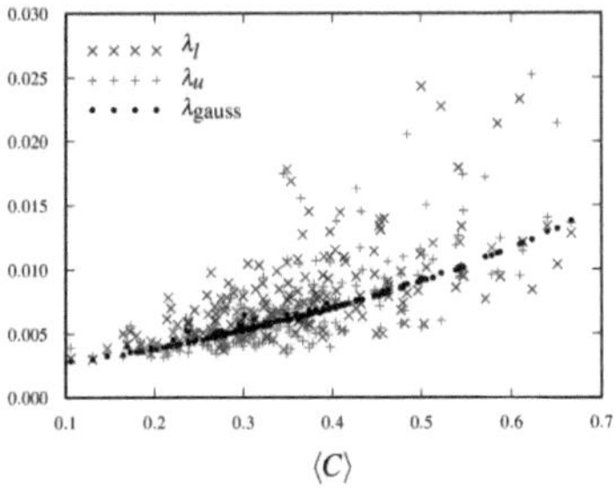

(i) $\Delta t = 120\text{min}, \alpha = 0.02$

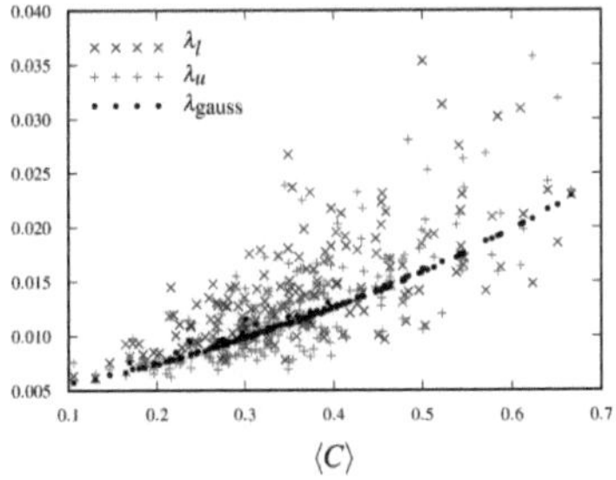

(j) $\Delta t = 120\text{min}, \alpha = 0.04$

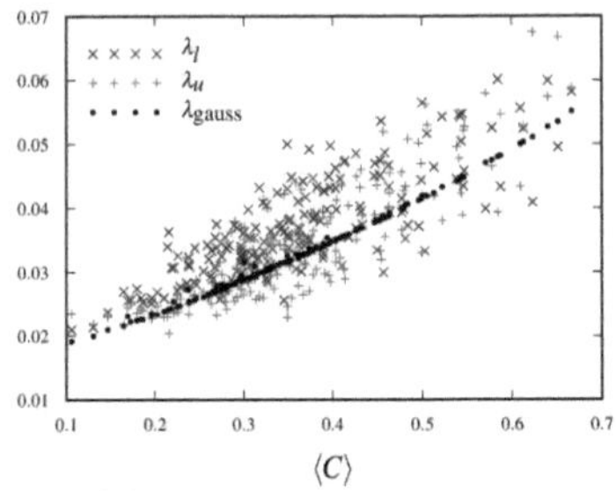

(k) $\Delta t = 120\text{min}, \alpha = 0.1$

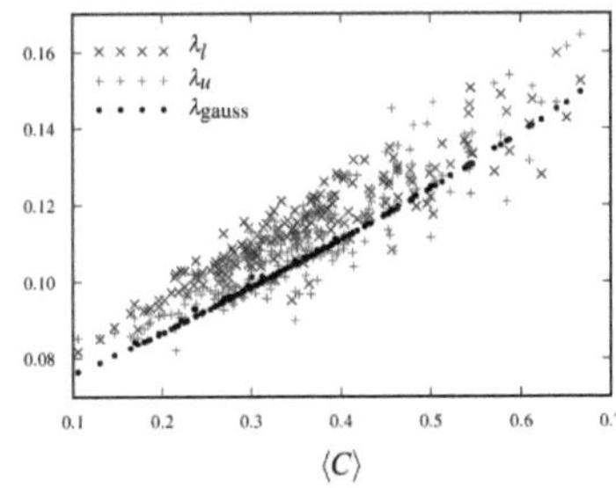

(l) $\Delta t = 120\text{min}, \alpha = 0.25$

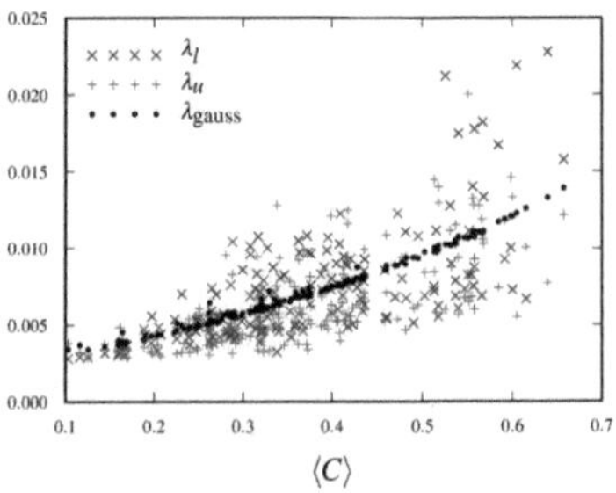

(m) $\Delta t = 240\text{min}, \alpha = 0.02$

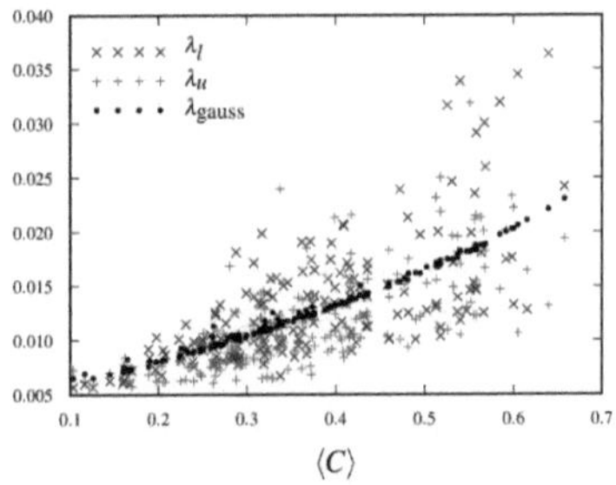

(n) $\Delta t = 240\text{min}, \alpha = 0.04$

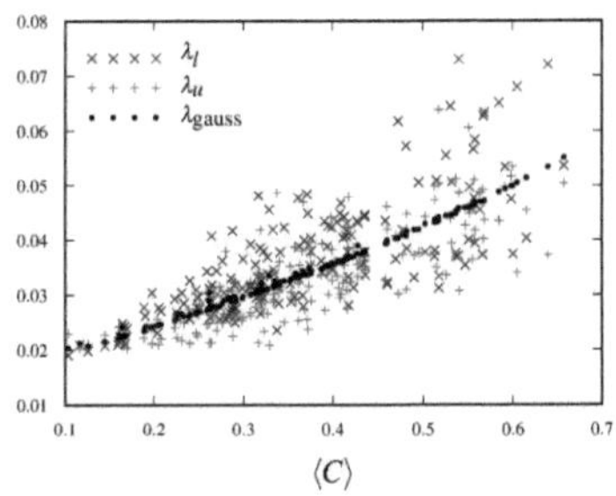

(o) $\Delta t = 240\text{min}, \alpha = 0.1$

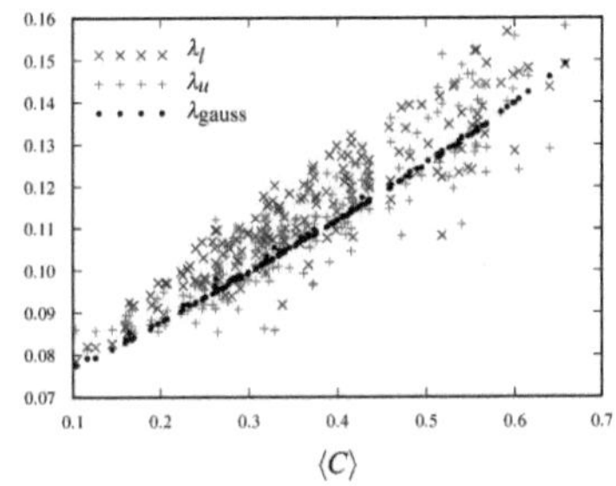

(p) $\Delta t = 240\text{min}, \alpha = 0.25$

Figure 2.14: (continued)

The relation between the market's average correlation level and the tail dependence appears to be almost linear. For small return intervals, such as $\Delta t = 30$min and 60min, the tail dependence has a tendency to be stronger than in the Gaussian case. For small quantiles, such as $\alpha = 2\%$ and 4%, there are many cases where this linear relation does not hold. There are many outliers that feature a much stronger tail dependence than in the Gaussian case. On larger return intervals, the tail dependence becomes more and more similar to the Gaussian case, which is consistent with studies of the marginal distributions [20]. Here, the lower tail dependence is significantly higher than the upper tail dependence. This underlines the unsuitability of the Gaussian copula for the estimation of *correlated* extreme events. This is a key ingredient to the estimation of financial risk [15, 31, 81, 82].

2.2.4 Summary

In a large-scale empirical study of the S&P 500 stock's copula, we uncovered important features of the statistical dependence structure. This gives the opportunity to isolate the statistical dependence structure from features of the marginal probability distributions, such as heavy tails. In general, the overall average pairwise copula of the 4-year period feature stronger tails than the Gaussian copula. Extreme events are much more correlated than assumed by a linear correlation. Moreover, the empirical copula indicates the presence of anti-correlated extreme events. Despite the large differences between the Gaussian distribution and the distribution of high frequency returns, the dependency structure in the central part of the distributions is quite similar. This explains why techniques to reduce risk that involve correlations work well in "quiet times", as these correspond to the center region of the copula. It also provides insights into the causes for the frequent failures of theses approaches during stock market crashes. The probability of simultaneous extreme events, both in correlated and anti-correlated manner, is underestimated.

In a time-dependent study, where we calculated the empirical copula in the resolution of 2-weeks, we showed that the Gaussian copula, in particular, systematically underestimates the tail dependence. The market reacts especially sensitive to large negative returns resulting in a collective downward motion. The evolution of the copula in the 4-year period discloses a strong relation between the market's average correlation level and the tail dependence. For large return intervals of 4 hours and in the center region of the distribution, the Gaussian copula describes the situation fairly well.

But when using smaller return intervals or estimating the tail regions, the fluctuations in the correlation/tail-dependence relation become very strong. This enables one to *dynamically* estimate the degree of error involved in the usage of correlation coefficients. It might be difficult to construct an analytically defined copula that fits the empirical data in all respects. However, a first step towards the reduction of risk is not only to know that correlated extreme events are underestimated, but also to be able to evaluate to which extent they are underestimated, given the market's overall correlation level.

3 Distorted Financial Correlations: The Epps Effect

The Epps effect describes the decrease of correlation estimates in financial data towards smaller return (or sampling-) intervals. This behavior has been of interest since Epps discovered this phenomenon in 1979 [89]. Since then, this behavior was found in data of different stock exchanges [90–93] and foreign exchange markets [94, 95]. An example for the Epps effect in empirical data is presented in Fig. 3.1. Here, the correlation declines for return intervals Δt smaller than five minutes.

Many economists as well as physicists addressed this phenomenon, because a precise calculation of correlations is of major importance for the estimation of financial risk [15, 27, 31, 96]. While the physicists' approach is often to construct a model, which offers an explanation for this phenomenon, the standard economy approach is to work on estimators with the aim to suppress the Epps effect.

Recently, Hayashi and Yoshida introduced an estimator [97], only involving returns whose time intervals are overlapping. Hence, it deals with the asynchrony of time series as a cause for the Epps effect. Subsequently, Voev and Lunde [98] demonstrated that this estimator can be biased in the presence of noise and proposed a bias correction. Griffin and Oomen [99] extended the estimator of Hayashi and Yoshida by adjustments for lagged correlations. The work of Tóth and Kertész [100] also deals with the phenomenon of lagged correlations. They introduce a model that is based on the decomposition of cross-correlations.

A very similar approach to the estimator of Hayashi and Yoshida, but on a completely different topic is the *discrete correlation function* in astrophysics which was introduced already in 1988 by Edelson and Krolik [101]. Other approaches to estimate correlations involve *Previous-Tick-Estimators* [102, 103] or realized kernel functions [104]. The term "previous-tick" simply refers to the fact that at a given time often no current price information exists. Thus, the last traded price, the so-called previous tick, is used.

The recent study of Zhang [103] shows that common previous-tick-estimators are biased. They consequently provide an optimal sampling fre-

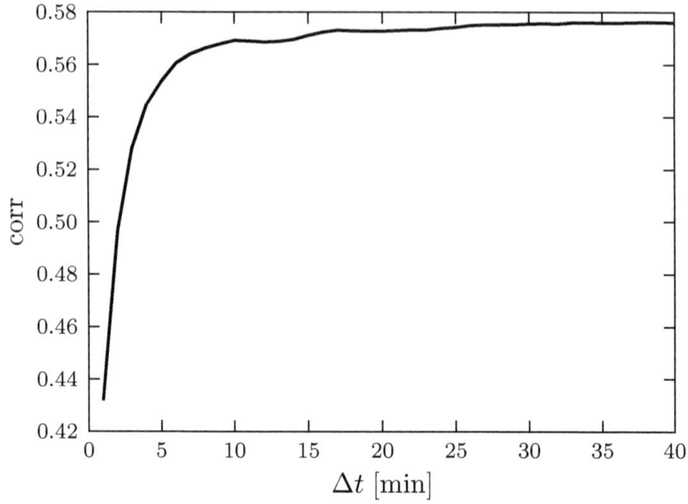

Figure 3.1: A typical Epps effect. The Fig. illustrates the Pearson correlation coefficient between Dominion Resources, Inc. (D) and Xcel Energy (XEL) in 2007 for various return intervals Δt.

quency of returns in order to suppress the Epps effect. Barndorff-Nielson et. al. [104, 105] examine high frequency correlations and propose multivariate realized kernels to improve the estimation of correlations. An extensive study of microscopic causes leading to the Epps effect has been performed by Renò [106].

Certainly many mechanisms contribute to the Epps effect. We can separate these mechanisms into two classes. First, statistical effects that originate from the Markovian features of the time series, e.g., biased estimates and second, non-Markovian causes. The latter are represented by features that are not of purely statistical origin, such as trading strategies and static lead-lag effects, for example due to different time-zones.

We demonstrate that there are two major causes of purely statistical origin. The aim is not to develop a complete description of the Epps effect. It is rather to identify statistical causes that can be compensated *directly*, without the requirement of adjusting parameters, model calibrations or an optimal sampling frequency, etc., which is typically required by other compensation methods for the Epps effect [93, 98, 99, 102–104].

The two major causes we identify are the asynchrony of the time series and the impact of the decimalization by the tick-size. The tick-size is the minimum price change of a stock's price. It is evident that the impact of the asynchrony – stock prices are not synchronized – grows at small return intervals. The asynchronous lags, usually in the range of seconds to minutes, can be neglected when calculating daily returns. We also expect a larger impact of the tick-size on small return intervals. Small return intervals are usually accompanied with small price changes. If these price changes are of the order of the tick-size, we expect a distortion of the correlation coefficients.

For both causes, we first introduce a simple model, which offers an explanation for this statistical part. Second, based on that model, we present an estimator, with which these effects can be compensated. Finally, we quantify the impact of this phenomenon on the Epps effect in recent empirical data and show that it can be a major cause for the Epps effect, especially when looking at less frequently traded securities.

This chapter is organized as follows[1]. In section 3.1, we discuss the impact of asynchronous time series on financial correlations and develop a compensation methods. The impact of the tick-size is taken into account in section 3.2. We combine both findings in section 3.3.

To evaluate the developed compensation methods, we set up a model that features asynchronous and discretized time series. Within the model set up, a decay of correlations towards smaller return intervals is observed which is discussed in section 3.4. Hence, the Epps effect can be reproduced within this model. Subsequently, the compensation methods developed in the preceding sections are validated within the model.

In section 3.5, this method is applied to recent empirical data to estimate the impact of the observed effect on the Epps effect. The results are discussed in section 3.6.

3.1 Asynchronous Time Series

We begin with demonstrating how the asynchrony of time series contributes to the Epps effect. By "asynchrony" we refer to time series that feature an arbitrary lag for a given point in time but the average lag is zero. The asynchrony is simply due to the non-synchronous pricing of stocks.

[1]For details see Refs. [1, 2].

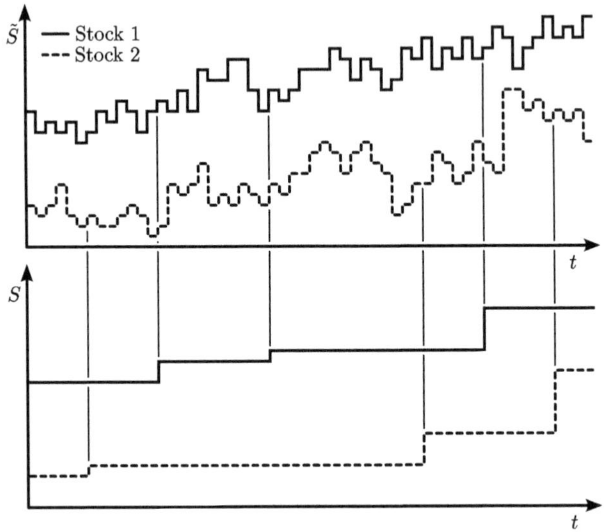

Figure 3.2: Illustration of the model for asynchronous trading times of two stocks. Shown above are the prices $\tilde{S}$ on the underlying timescale. The "sampling" of theses prices $\tilde{S}$ to prices S on simulated trading times are shown below.

The central assumption of this model is the existence of an underlying non-lagged time series of prices. The assumption of a finer [100] or even continuous [97, 105, 106] underlying timescale is a common approach in the estimation of correlations. This approach is also intuitive, as, e.g., most stocks are traded at several stock exchanges simultaneously.

3.1.1 Compensating the Correlation Coefficient for Asynchronous Effects

The basic idea of this approach is the following: Due to the asynchrony, each term of the correlation coefficient can be divided into a part which contributes to the correlation and a part which is uncorrelated and therefore lowers the correlation coefficient.

According to the model assumption, the price change during Δt is based on price changes on an underlying "microscopic" timescale. Thus, the return

$r_{\Delta t}$ can also be expressed as a sum of the underlying returns $\tilde{r}$,

$$r^{(i)}_{\Delta t}(t) = \sum_{j=0}^{N^{(i)}_{\Delta t}(t)} \tilde{r}^{(i)}(\gamma^{(i)}(t) + j\Delta\tilde{t}) \ . \tag{3.1}$$

Here $\tilde{r}^{(i)}(t_i)$ is the return related to $S(t)$ on the underlying time scale of non-overlapping intervals $\Delta\tilde{t}$ (e.g. 1 second) given by

$$\tilde{r}^{(i)}(t + j\Delta\tilde{t}) = \frac{\tilde{S}(t + (j+1)\Delta\tilde{t}) - \tilde{S}(t + j\Delta\tilde{t})}{S(t)} \ . \tag{3.2}$$

The quantity $\gamma^{(i)}(t)$ in Eq. (3.1) represents the time of the last trade of the i-th stock at time t,

$$\gamma^{(i)}(t) = \max(t^{(i)}_{\text{trade}})\Big|_{t^{(i)}_{\text{trade}} \leq t} \ . \tag{3.3}$$

When calculating the return of the interval $[t, t{+}\Delta t]$ of two stocks, the actual price at t and $t + \Delta t$ generally originates from the past, more precisely at the times $\gamma^{(1)}(t)$, $\gamma^{(2)}(t)$ and $\gamma^{(1)}(t + \Delta t)$, $\gamma^{(2)}(t + \Delta t)$. These intervals can be smaller or larger than the initially chosen return interval. When considering the returns of two stocks within the same interval, two effective return intervals are obtained that are in most cases not equal in length, start-point and end-point. Therefore only a fraction of the underlying prices processed by the return are correlated. The number of terms $N^{(i)}_{\Delta t}$ of the sum in Eq. (3.1) is given by

$$N^{(i)}_{\Delta t}(t) = \frac{(\gamma^{(i)}(t + \Delta t) - \gamma^{(i)}(t))}{\Delta\tilde{t}} \ . \tag{3.4}$$

For the sake of a simpler notation, we normalize the returns to zero mean and unit variance and indicate them as g and $\tilde{g}$,

$$g^{(i)}_{\Delta t}(t) = \frac{r^{(i)}_{\Delta t}(t) - \langle r^{(i)}_{\Delta t}\rangle}{\sigma^{(i)}_{\Delta t}} \tag{3.5}$$

$$\tilde{g}^{(i)}(t) = \frac{\tilde{r}^{(i)}(t) - \langle \tilde{r}^{(i)}\rangle}{\tilde{\sigma}^{(i)}} \ . \tag{3.6}$$

To derive the relationship between the normalized returns $\tilde{g}^{(i)}$ on the underlying timescale and $\tilde{g}^{(i)}$ the macroscopic timescale, we start with inserting the return, expressed through the sum of small "returns" on underlying time series,

$$r_{\Delta t}^{(i)}(t) = \sum_{j=0}^{N_{\Delta t}^{(i)}(t)} \tilde{r}^{(i)}(\gamma^{(i)}(t) + j\Delta\tilde{t}) \ , \tag{3.7}$$

in Eq. (3.5). This leads to

$$g_{\Delta t}^{(i)}(t) = \frac{\sum\limits_{j=0}^{N_{\Delta t}^{(i)}(t)} \left(\tilde{r}^{(i)}(\gamma^{(i)}(t) + j\Delta\tilde{t})\right) - \langle r_{\Delta t}^{(i)} \rangle}{\sqrt{\mathrm{var}(r_{\Delta t}^{(i)})}} \tag{3.8}$$

$$= \frac{\sqrt{\mathrm{var}(\tilde{r}^{(i)})}}{\sqrt{\mathrm{var}(r_{\Delta t}^{(i)})}} \times \sum_{j=0}^{N_{\Delta t}^{(i)}(t)} \left(\tilde{g}^{(i)}(\gamma^{(i)}(t) + j\Delta\tilde{t})\right) - \left\langle r_{\Delta t}^{(i)} \right\rangle + N_{\Delta t}^{(i)}(t) \left\langle \tilde{r}^{(i)} \right\rangle \ . \tag{3.9}$$

In Eq. (3.9), Eq. (3.6) was used to express the underlying returns $\tilde{r}^{(i)}$. As $\tilde{r}^{(i)}$ and $N_{\Delta t}^{(i)}$ are uncorrelated and the mean values and variances are additive, we have

$$\left\langle r_{\Delta t}^{(i)} \right\rangle = \left\langle N_{\Delta t}^{(i)} \right\rangle \left\langle \tilde{r}^{(i)} \right\rangle \tag{3.10}$$

$$\mathrm{var}\left(r_{\Delta t}^{(i)}\right) = \left\langle N_{\Delta t}^{(i)} \right\rangle \mathrm{var}\left(\tilde{r}^{(i)}\right) \ . \tag{3.11}$$

Therefore, we obtain by insertion into Eq. (3.9)

$$g_{\Delta t}^{(i)}(t) = \frac{1}{\sqrt{\left\langle N_{\Delta t}^{(i)} \right\rangle}} \sum_{j=0}^{N_{\Delta t}^{(i)}(t))} \tilde{g}^{(i)}(\gamma^{(i)}(t) + j\Delta\tilde{t}) - \frac{\langle \tilde{r}^{(i)} \rangle \left(\left\langle N_{\Delta t}^{(i)} \right\rangle - N_{\Delta t}^{(i)}(t) \right)}{\sqrt{\mathrm{var}(r_{\Delta t}^{(i)})}} \ . \tag{3.12}$$

As the average time interval per return converges to Δt, the mean number of underlying price changes $\left\langle N_{\Delta t}^{(i)} \right\rangle$ is given by $\Delta t / \Delta \tilde{t}$. Thus, we arrive at

$$g_{\Delta t}^{(i)}(t) = \sqrt{\frac{\Delta \tilde{t}}{\Delta t}} \sum_{j=0}^{N_{\Delta t}^{(i)}(t))} \tilde{g}^{(i)}(\gamma^{(i)}(t) + j\Delta \tilde{t}) - \frac{\langle \tilde{r}^{(i)} \rangle \left(\frac{\Delta t}{\Delta \tilde{t}} - N_{\Delta t}^{(i)}(t) \right)}{\sqrt{\mathrm{var}(r_{\Delta t}^{(i)})}} . \quad (3.13)$$

When using normalized returns, the correlation coefficient of two return time series $r_{\Delta t}^{(1)}$ and $r_{\Delta t}^{(2)}$ (see Eq. (1.8) simplifies to

$$\mathrm{corr}(r_{\Delta t}^{(1)}, r_{\Delta t}^{(2)}) = \mathrm{corr}(g_{\Delta t}^{(1)}, g_{\Delta t}^{(2)}) = \frac{1}{T} \sum_{j=0}^{T} g_{\Delta t}^{(1)}(t_j) g_{\Delta t}^{(2)}(t_j) . \quad (3.14)$$

As the mean value over T of the second term from Eq. (3.13) is equal to zero, we obtain in terms of the underlying time series

$$\begin{aligned} \mathrm{corr}(r_{\Delta t}^{(1)}, r_{\Delta t}^{(2)}) = \frac{1}{T} \sum_{j=0}^{T} \Bigg(& \sum_{k=0}^{N_{\Delta t}^{(1)}(t_j)} \tilde{g}^{(1)}(\gamma^{(1)}(t_j) + k\Delta \tilde{t}) \\ & \times \sum_{l=0}^{N_{\Delta t}^{(2)}(t_j)} \tilde{g}^{(2)}(\gamma^{(2)}(t_j) + l\Delta \tilde{t}) \Bigg) \frac{\Delta \tilde{t}}{\Delta t} . \end{aligned} \quad (3.15)$$

Fig. 3.3 illustrates that only a subset of the underlying prices $\tilde{S}$ of two prices S share an overlapping time-interval. Because of this “overlap” only a certain amount $\bar{N}_{\Delta t}(t)$ of the underlying returns is correlated, namely

$$\bar{N}_{\Delta t}(t) = \frac{\Delta t_{\mathrm{o}}(t)}{\Delta \tilde{t}} , \quad (3.16)$$

with $\Delta t_{\mathrm{o}}(t)$ being the time interval of the actual overlap, in which both stocks have synchronous prices

$$\Delta t_{\mathrm{o}}(t) = \min(\gamma^{(1)}(t + \Delta t), \gamma^{(2)}(t + \Delta t)) - \max(\gamma^{(1)}(t), \gamma^{(2)}(t)) . \quad (3.17)$$

Each sum can be split up into $N_{\Delta t}^{(i)} - \bar{N}$ terms that are uncorrelated and $\bar{N}$

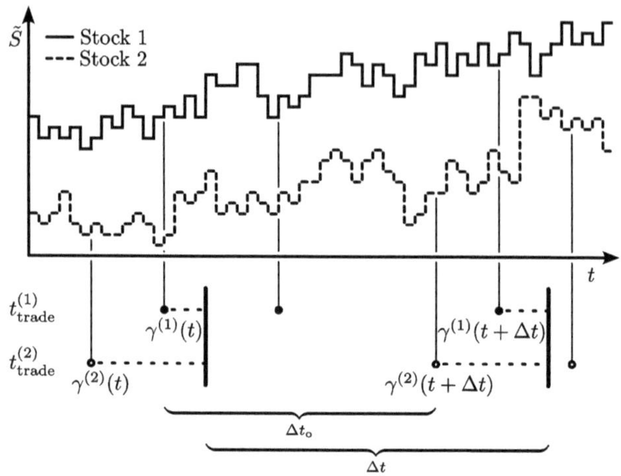

Figure 3.3: Illustration of the overlap Δt_{o} that originates from asynchronous price information.

that are correlated. Thus, Eq. (3.15) can be written as

$$\operatorname{corr}(r^{(1)}_{\Delta t}, r^{(2)}_{\Delta t}) = \frac{1}{T}\sum_{j=0}^{T}\left(\left(\underbrace{\sum_{k=\bar{N}_{\Delta t}(t_j)+1}^{N^{(1)}_{\Delta t}(t_j)-\bar{N}_{\Delta t}(t_j)} \tilde{g}^{(1)}(t_k)}_{\text{async.}} + \underbrace{\sum_{\bar{k}=0}^{\bar{N}_{\Delta t}(t_j)} \tilde{g}^{(1)}(t_{\bar{k}})}_{\text{sync.}}\right) \times \left(\underbrace{\sum_{l=\bar{N}_{\Delta t}(t_j)+1}^{N^{(2)}_{\Delta t}(t_j)-\bar{N}_{\Delta t}(t_j)} \tilde{g}^{(2)}(t_l)}_{\text{async.}} + \underbrace{\sum_{\bar{l}=0}^{\bar{N}_{\Delta t}(t_j)} \tilde{g}^{(2)}(t_{\bar{l}})}_{\text{sync.}}\right)\frac{\Delta\tilde{t}}{\Delta t}\right), \tag{3.18}$$

where only the sums of synchronous returns are correlated among each other. In this notation, the underlying time series is indexed as $[\tilde{r}^{(i)}(t_0), \tilde{r}^{(i)}(t_1), \ldots, \tilde{r}^{(i)}(t_{N^{(i)}_{\Delta t}})]$, where the returns from t_0 to $t_{\bar{N}_{\Delta t}}$ are corresponding to the overlap. When expanding the product, the non-correlated returns converge to zero due to the outer average

$$\mathrm{corr}(r^{(1)}_{\Delta t}, r^{(2)}_{\Delta t}) = \frac{1}{T}\sum_{j=0}^{T}\left(\left(\underbrace{\sum_{k=0}^{\bar{N}_{\Delta t}(t_j)} \tilde{g}^{(1)}(t_k)\tilde{g}^{(2)}(t_k)}_{\bar{N}_{\Delta t}(t)\mathrm{corr}_{t_j}(\vec{\tilde{r}}_1,\vec{\tilde{r}}_2)} + \underbrace{\ldots}_{0}\right)\frac{\Delta\tilde{t}}{\Delta t}\right)$$
$$= \frac{1}{T}\sum_{j=0}^{T}\mathrm{corr}_{t_j}(\tilde{g}^{(1)}, \tilde{g}^{(2)})\frac{\bar{N}_{\Delta t}(t)\Delta\tilde{t}}{\Delta t}$$
$$= \frac{1}{T}\sum_{j=0}^{T}\mathrm{corr}_{t_j}(\tilde{g}^{(1)}, \tilde{g}^{(2)})\frac{\Delta t_{\mathrm{o}}(t_j)}{\Delta t}\,, \tag{3.19}$$

where corr_t represents the correlation of the underlying returns corresponding to the interval $[t, t + \Delta t]$.

$\Delta t_{\mathrm{o}}(t)/\Delta t$ is the fractional overlap of the corresponding return interval. This fractional overlap does not depend on the actual timescale of the underlying time series. As Eq. (3.19) clearly shows, the correlation coefficient of the synchronous part of the return time series is multiplied by the fractional overlap. Hence, this effect can be compensated by

$$\widehat{\mathrm{corr}}_{\mathrm{async}}(r^{(1)}_{\Delta t}, r^{(2)}_{\Delta t}) = \frac{1}{T}\sum_{j=0}^{T} g^{(1)}_{\Delta t}(t_j) g^{(2)}_{\Delta t}(t_j)\frac{\Delta t}{\Delta t_{\mathrm{o}}(t_j)}\,, \tag{3.20}$$

which is the final result for the asynchrony compensation.

To review – Initially, we made the assumption of an underlying time series of prices, which is correlated and which exists on a smaller time scale. Eq. (3.20) does no longer depend on the time scale of the hypothetical underlying time series. Neither does it depend on the actual prices on the underlying time series. Hence, the only necessary assumption is that there exists underlying *information*, which is correlated on a finer time scale. This is an important finding, since the synchronization of returns from international stock exchanges is a highly non-trivial problem [107].

3.2 Impact of the Tick-Size

Observations on financial data on very small time scales or small amplitudes are usually referred to as market microstructure [70]. The following demonstrates that this microstructure has a large impact on the correlation estimation and can also alter the tail behavior of the return distribution compared to the underlying price change distribution. Before we turn to the Epps effect, we disclose a relation between the tail behavior of each microstructure return distribution for a fixed price change $\Delta S_{\Delta t}$ and the overall return distribution. For this purpose, we decompose the set of returns according to the absolute price changes and disclose its microstructure.

The *tick-size* or *minimum tick*, plays an important role in quantitative finance. All raw price information is discretized by the tick-size. Historically, the tick-size of most securities has been consecutively reduced resulting in tick-sizes of 1/100th. This process is often referred to as decimalization [108]. One reason for it was to aim at enhanced market efficiency. In principle, small tick-sizes allow for a faster clearing of market arbitrage. Nonetheless, it is controversial whether a smaller tick-size generally improves the market quality [109–112], e.g., in view of the fact that a larger tick-size ensures liquidity [113]. Furthermore, a recent study indicates that in some cases only a fraction of the theoretically possible prices are used. Hence, prices cluster at certain multiples of the tick-size resulting in an effective tick-size [114].

However, a large tick-size can lead to erroneous data in financial indices due to rounding errors [115]. The actual tick-size for stocks is typically USD 0.01. This is the case for instance on the New York Stock Exchange (NYSE) and the National Association of Securities Dealers Automated Quotations (NASDAQ). However, some securities such as U.S. Government securities are still quoted in 1/32nds of a dollar.

The tick-size certainly affects many fields in quantitative finance. In this section we want to focus on its impact on two of the most important observables: financial returns and financial correlations. These elementary values are of particular importance for many applications, for example portfolio optimization [52, 116] and risk management [15].

In section 3.2.1, the influence of the tick-size on the microstructure of financial return distributions is studied. Subsequently, impact of the tick-size on the calculation on the Epps Effect. As the identified mechanism is solely of statistical origin, we are able to develop a method for compensating this distortion as well, as discussed in section 3.2.2.

3.2.1 Return Microstructure

A *financial return* describes the relative price change of a security between two points in time. As introduced in section 1.2.1, the arithmetic return is defined as

$$r_{\Delta t}(t) = \frac{\Delta S_{\Delta t}(t)}{S(t)} \; . \tag{3.21}$$

As the price change $\Delta S_{\Delta t}$ can only take values that are multiples of the tick-size q, its histogram consists of equally spaced peaks as shown in Fig. 3.4. In other words, the distribution of $\Delta S_{\Delta t}$ is discretized. At first glance, it is conceivable that the transition from absolute price changes $\Delta S_{\Delta t}$ to relative price changes $r_{\Delta t}$ removes this discretization from the distribution, since the returns are almost continuously distributed, as Fig. 3.5 illustrates. However, a closer look at the center of the distribution in Fig. 3.6 reveals that the discretization effects are still visible. Despite its non-visibility, the discretization affects returns on any interval. We discuss this point more detailed in sections 3.2.2 to 3.2.5.

For an analytical description of this discretization, we introduce the set of all returns

$$R_{\Delta t} = \left\{ \frac{\Delta S_{\Delta t}(t)}{S(t)} \,\middle|\, \Delta S_{\Delta t}(t) \in [N_{-}q, (N_{-}+1)q, \ldots, (N_{+}-1)q, N_{+}q] \right\} \; , \tag{3.22}$$

where $N_{-}q$ defines the lower and $N_{+}q$ the upper bound of the price change distribution that is discretized by the tick-size q.

The set of all returns $R_{\Delta t}$ can be separated into subsets for each price change $\Delta S_{\Delta t}$,

$$R_{\Delta t} = \bigcup_{n=N_{-}}^{N_{+}} R_{\Delta t}^{(n)} \; , \tag{3.23}$$

with

$$R_{\Delta t}^{(n)} = \left\{ \frac{\Delta S_{\Delta t}(t)}{Y^{(n)}(t)} \,\middle|\, \Delta S_{\Delta t}(t) = nq \right\} \; . \tag{3.24}$$

$Y^{(n)}$ in the denominator refers to the subset of starting prices S that increase (or decrease) by nq_S in the interval Δt. Therefore, $R_{\Delta t}^{(n)}$ represents the returns that are based on the price change nq. Evidently, $R_{\Delta t}^{(n)}$ is bounded by

$$\min(R_{\Delta t}^{(n)}) = \frac{nq}{\max(Y^{(n)})} \quad , \quad \max(R_{\Delta t}^{(n)}) = \frac{nq}{\min(Y^{(n)})} \; . \tag{3.25}$$

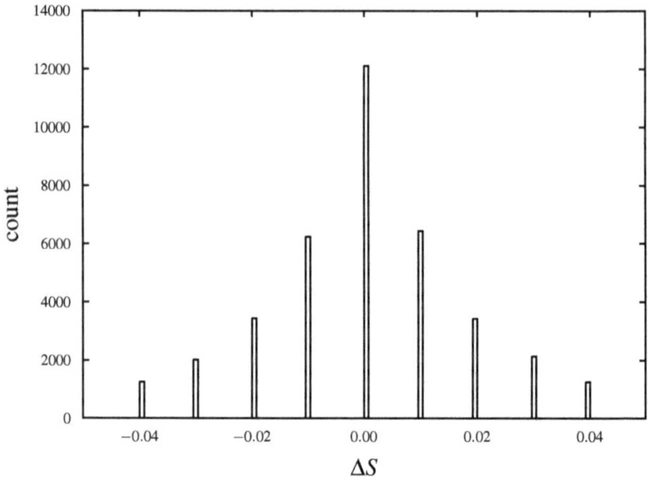

Figure 3.4: Center of the 1-minute price change distribution from the Apollo Group Inc. (APOL) share in the first half of 2007.

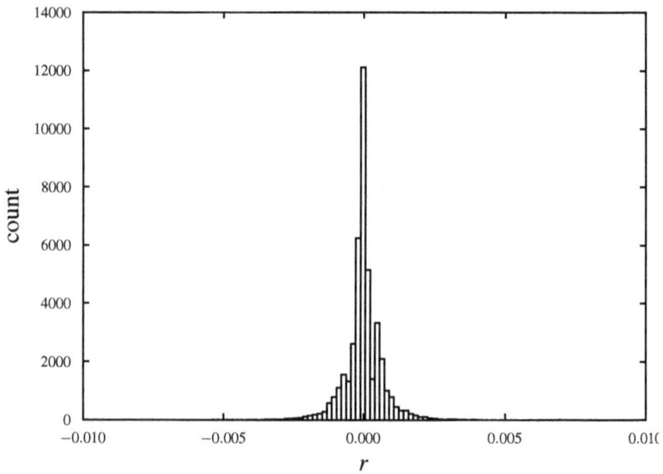

Figure 3.5: 1-minute return distribution from the Apollo Group Inc. (APOL) share in the first half of 2007.

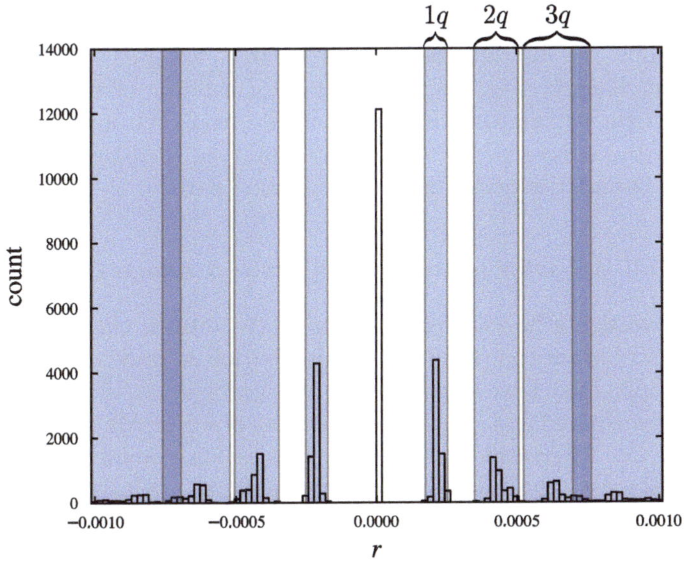

Figure 3.6: Center of the 1-minute return distribution from the Apollo Group Inc. (APOL) share in the first half of 2007 with the calculated bounds corresponding to a specific price change indicated by the blue regions. Darker shades of blue imply overlapping bounds.

In our study, empirical data from the TAQ database [43] of the New York Stock exchange (NYSE) indicate that the approximations $\max(Y^{(n)}) \approx \max(Y)$ and $\min(Y^{(n)}) \approx \min(Y)$ are legitimate for small $|n|$, where Y is the set of all prices in the observed period.

Therefore, the interval between minimum and maximum return on a specific price change

$$I(R^{(n)}) = \left[\min\left(R^{(n)}\right), \max\left(R^{(n)}\right)\right] \tag{3.26}$$

increases with $|n|$, while the distance d between their centers remains almost constant

$$\begin{aligned} d(R^{(n)}) &= \frac{q_S}{2}\left(\frac{1}{\min(Y^{(n)})} - \frac{1}{\max(Y^{(n)})}\right) \\ &\approx \frac{q_S}{2}\left(\frac{1}{\min(Y)} - \frac{1}{\max(Y)}\right) = \text{const.} \end{aligned} \tag{3.27}$$

Thus, the intervals $I(Y^{(n)})$ are increasingly overlapping for larger $|n|$. From this viewpoint the discretization is only "visible" for small $|n|$, that is, for small price changes. Fig. 3.6 illustrates the clustering of returns with an example where we compare the returns of the Apollo Group Inc. (APOL) share with the intervals $I(R^{(n)})$ calculated by Eqs. (3.26) and (3.27). The calculated boundaries match with the empirical data.

3.2.1.1 Tail Behavior of Return and Price Change Distributions

Now we investigate how the composition of the returns changes the shape of their distribution compared to the distribution of price changes. In the framework of a model, we generate price changes that are, in a first scenario, Gaussian distributed and, in a second scenario, power-law distributed with a given tick-size. Afterwards, we calculate returns using uniformly distributed price values within the minimum and maximum price, S_{min} and S_{max} (analogously to Figs. 3.6 and 3.5). In this manner, we generate a discrete price change distribution with a specific shape and then divide each set of equal price changes by uniformly distributed prices. The price distributions are generated individually for each subset.

To compare the shape of the obtained return distribution with the shape, which we have chosen for the price change distribution, we normalize the distributions to zero mean and unit variance

$$g_{\Delta t}^{(i)}(t) = \frac{r_{\Delta t}^{(i)}(t) - \langle r_{\Delta t}^{(i)} \rangle}{\sigma_{\Delta t}^{(i)}} \tag{3.28}$$

$$\Delta \hat{S}_{\Delta t}^{(i)}(t) = \frac{\Delta S_{\Delta t}^{(i)}(t) - \langle \Delta S_{\Delta t}^{(i)} \rangle}{\sigma_{\Delta t}^{(i)}} \; . \tag{3.29}$$

The results of this simple setup indicate that neither the tick-size nor the width of the price change distribution or the absolute sizes of S_{min} and S_{max} have an effect on the shape of the obtained return distribution. Only the microstructure of its center is affected, as discussed in the previous section. In general, the return distribution acquires stronger tails compared to the price change distribution. Surprisingly, the shape-change of the distribution only depends on the ratio of the minimum and maximum price.

Fig. 3.7 shows the corresponding distributions for Gaussian and power-law distributed prices and for various price ranges. It turns out that the influence on the tail behavior is much stronger for a Gaussian price change

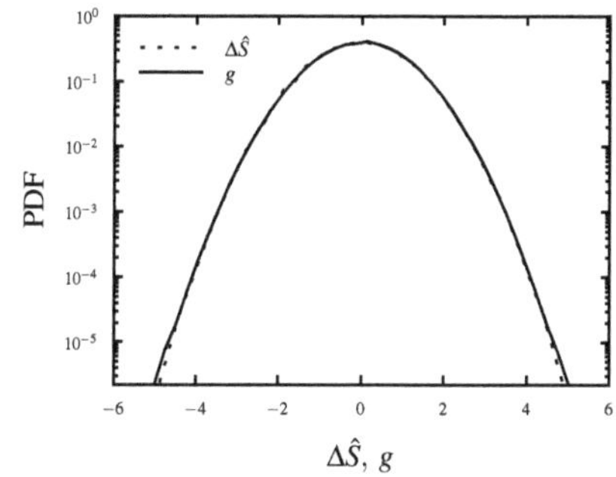

(a) Gaussian ΔS, $S_{\text{max}}/S_{\text{min}} = 1.1$

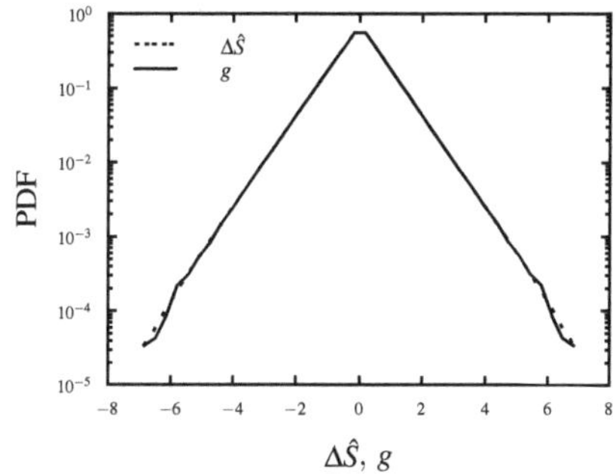

(b) Power-law ΔS, $S_{\text{max}}/S_{\text{min}} = 1.1$

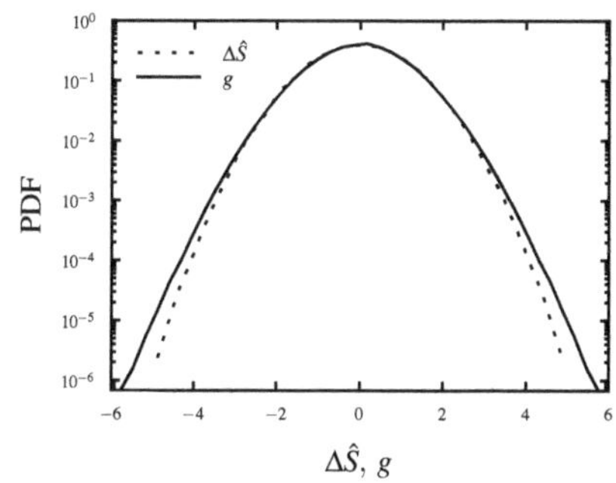

(c) Gaussian ΔS, $S_{\text{max}}/S_{\text{min}} = 1.5$

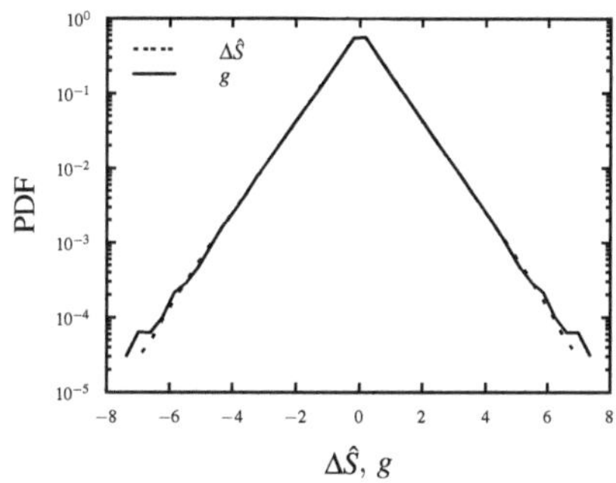

(d) Power-law ΔS, $S_{\text{max}}/S_{\text{min}} = 1.5$

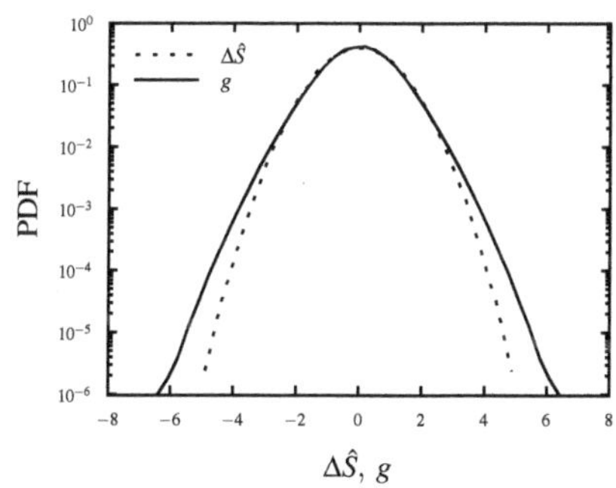

(e) Gaussian ΔS, $S_{\text{max}}/S_{\text{min}} = 2.0$

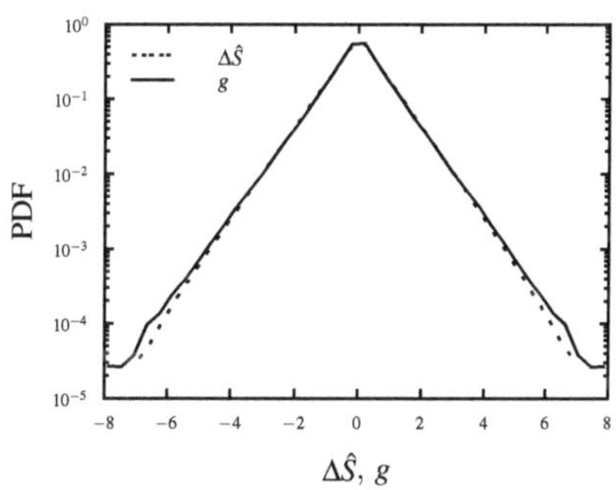

(f) Power-law ΔS, $S_{\text{max}}/S_{\text{min}} = 2.0$

Figure 3.7: Comparison of the distributions of normalized price changes $\Delta\hat{S}$ and normalized returns g on different price ranges $S_{\text{max}}/S_{\text{min}}$ using Gaussian distributed price changes (a,c,e) and power-law distributed price changes (b,e,f). All calculations were preformed using a standard deviation of 60 tick-sizes.

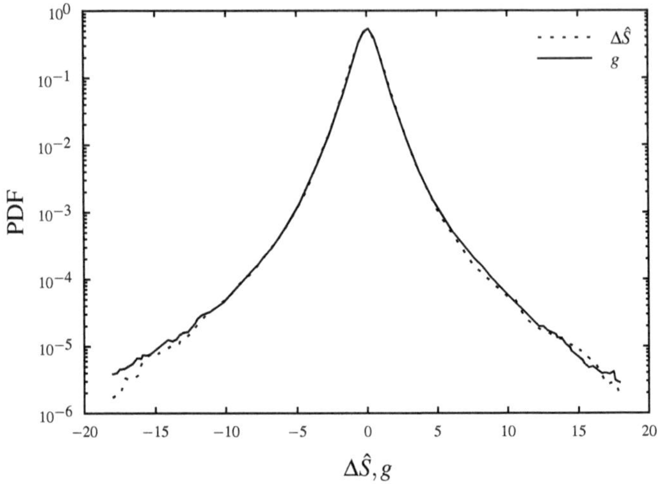

(a) $\Delta t = 5$ min

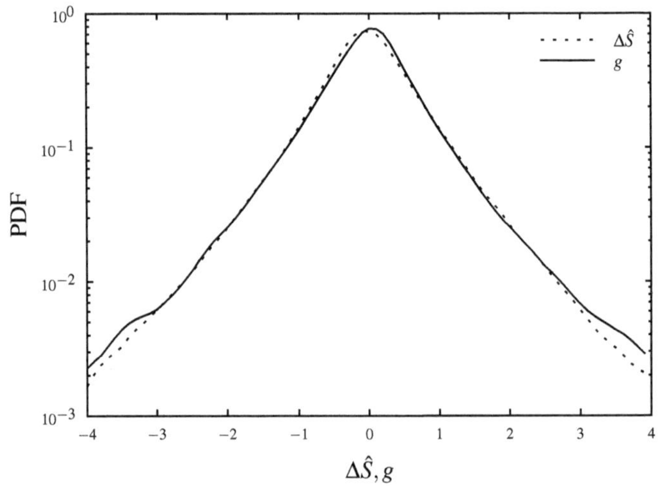

(b) $\Delta t = 1$ d

Figure 3.8: Change in the tail behavior of the distribution of normalized returns g compared to the underlying normalized price changes $\Delta\hat{S}$.

distribution. For a power-law price change distribution, the return distribution retains approximately the same power-law shape, except for the far tails and the slight sharpening of the center region.

Certainly, the assumption of uniformly distributed prices on each price change is a rough approximation within this simple setup. In the market, there can be a strong relation between ΔS and S, which leads to a shape retaining of the price change distribution to the return distribution. This is because the prices which undergo a very large price change during the interval Δt can be much more sparsely distributed than prices which change only slightly. Furthermore, the price range is usually not large in a period of time, in which the price distribution is approximately uniform. In view of this and under the assumption of power-law distributed price changes, the situation in Figs. 3.7b and 3.7d may describe most stocks suitably. Put differently, the shape of the return distribution is almost retaining the shape of the price change distribution in most cases.

However, if the price of a stock covers a large range in a relatively short period of time, we actually can observe a change in the tail behavior. This is illustrated in Fig. 3.8 for an ensemble of 50 stocks taken from the S&P 500 index (see appendix A.3) using return intervals of 5 minutes and 1 day. The stocks have been chosen to provide the highest ratio between their mean price and its standard deviation. Although the stock ensemble shows the expected behavior, it is difficult to make an accurate statement regarding the far tails, as these events are very rare, even within this statistical ensemble.

3.2.2 Impact of the Tick-Size on Financial Correlations

We now turn to the impact of the tick-size on the calculation of correlations and analyze the impact on the Epps effect. Financial correlations are an important measure in economics. The knowledge of precise correlations is essential for quantifying and minimizing financial risk. As we will show, the discreteness of stock quotes can distort the correlation estimates.

It is a basic assumption in our approach that we can statistically describe the discreteness in market prices by a discretization of a hypothetical underlying continuous price. This is not to say that market prices actually result from a discretization process. Individual traders are well aware of the finite tick-size and may try to exploit it in their trading strategies. However, there is a large variety of trading strategies simultaneously acting on the market. These strategies involve a large scale of different investment horizons. Since the price formation results from the interaction of a large diversity of strategies, the price fluctuations on the level of the tick-size can be viewed as purely statistical. This is the basis for our modeling ansatz. Despite the interpolation of the price change distribution, neither parameter fixing nor calibration of the model is necessary.

A financial return is a compound observable value. Due to this fact, we develop the compensation method step by step. We start with turning to the distortion of the correlation coefficient of value-discretized time series in general in section 3.2.3. We develop a compensation for the discretization error in the correlation between financial (absolute) price changes in section 3.2.4. Eventually, we extend this formalism to financial returns in section 3.2.5. To simplify the notation, we omit the index Δt indicating the return interval in sections 3.2.4 and 3.2.5.

3.2.3 Estimating Correlations in Value-Discretized Time Series in General

Almost any time series of data is discretized. This can simply be caused by numerical reasons, such as a finite number of decimal places. But how can we measure the impact of the discretization or even compensate it? We demonstrate, that this can be achieved by a decomposition of the correlation coefficient and a estimation of the average discretization errors.

Let x_1 and x_1 be two time series which are correlated. Now, let us consider the time series $\bar{x}_1$ and $\bar{x}_2$ which are the discretized values of x_1 and x_2 with tick-sizes q_1 and q_2, respectively. Thus we have

$$x_1(t) = \bar{x}_1(t) + \vartheta^{(1)}(t) \tag{3.30}$$

$$x_2(t) = \bar{x}_2(t) + \vartheta^{(2)}(t) \ , \tag{3.31}$$

where $\vartheta^{(1)}(t)$ and $\vartheta^{(2)}(t)$ are the discretization errors. We assume the discretization errors as uniformly distributed in the intervals $]-q_1/2, q_1/2]$ and $]-q_2/2, q_2/2]$. This seems natural, as discretization is commonly caused by a rounding process. Using Eqs. (3.30) and (3.31) we can write the correlation coefficient as

$$\operatorname{corr}(x_1,x_2) = \frac{\langle(\bar{x}_1+\vartheta^{(1)})(\bar{x}_2+\vartheta^{(2)})\rangle - \left(\langle\bar{x}_1\rangle+\langle\vartheta^{(1)}\rangle\right)\left(\langle\bar{x}_2\rangle+\langle\vartheta^{(2)}\rangle\right)}{\sqrt{\operatorname{var}\left(\bar{x}_1+\vartheta^{(1)}\right)}\sqrt{\operatorname{var}\left(\bar{x}_2+\vartheta^{(2)}\right)}} \tag{3.32}$$

$$= \frac{\operatorname{cov}(\bar{x}_1,\bar{x}_2)+\operatorname{cov}\left(\bar{x}_1,\vartheta^{(2)}\right)+\operatorname{cov}\left(\bar{x}_2,\vartheta^{(1)}\right)+\operatorname{cov}\left(\vartheta^{(1)},\vartheta^{(2)}\right)}{\hat{\sigma}_{x_1}\hat{\sigma}_{x_2}} \tag{3.33}$$

with

$$\hat{\sigma}_{x_i} = \sqrt{\operatorname{var}(\bar{x}_i)+\operatorname{var}\left(\vartheta^{(i)}\right)+2\operatorname{cov}\left(\bar{x}_i,\vartheta^{(i)}\right)}\ . \tag{3.34}$$

Apart from the terms $\operatorname{cov}(\bar{x}_1,\bar{x}_2)$, $\operatorname{var}(\bar{x}_1)$ and $\operatorname{var}(\bar{x}_2)$ of expression (3.33), which can be calculated with the discretized data, all other terms are lost in the discretization process. However, these terms can be estimated when the distributions $\varrho_{\bar{x}_1}$ and $\varrho_{\bar{x}_2}$ of $\bar{x}_1$ and $\bar{x}_2$ are known, as we will demonstrate. The continuous distributions ϱ_{x_1} and ϱ_{x_2} can be obtained by interpolating the distributions of the discretized values. We assume these distributions in the following context to be normalized. Sometimes, the shape of the distribution for a certain process is known (e.g. Gaussian). Therefore, the interpolated distribution function can be determined by a fit of the distributions of $\bar{x}_1$ and $\bar{x}_2$.

If the shape of the distribution is unknown, an interpolation can be performed section by section using, e.g., polynomial or linear fits. The fitting processes cannot be performed in the traditional way by minimizing the difference of values from the discrete distribution and the desired fit function. Rather the discretization process needs to be included. This gains particular importance when the level of discretization is high and thus the distribution is discretized only with a small range of values.

As the value that is discretized to, e.g., x_1' can originate from region $x_1'-q_1/2$ to $x_1'+q_1/2$, the difference function f, which provides a measure for the residual between the fit and the empirical data is then given by

$$f_{x_1}(\varrho_{x_1},\varrho_{\bar{x}_1}) = \sum_{n=N_-}^{N_+}\left[\int\limits_{q_1(n-\frac{1}{2})}^{q_1(n+\frac{1}{2})}\varrho_{x_1}(z)\,dz - \varrho_{\bar{x}_1}(nq_1)\right] \tag{3.35}$$

for x_1 and analogously for x_2.

To compensate the overall discretization error, we first introduce the discretization errors that led to a certain discretized value. We refer to these errors as conditional discretization errors. They are defined as

$$\vartheta_n^{(1)} = x_1(\tilde{t}) - nq_1 \ , \ \tilde{t} \in \left\{ t \mid |x_1(t) - nq_1| \leq \frac{q_1}{2} \right\} \tag{3.36}$$

$$\vartheta_m^{(2)} = x_2(\tilde{t}) - mq_2 \ , \ \tilde{t} \in \left\{ t \mid |x_2(t) - mq_2| \leq \frac{q_2}{2} \right\} \tag{3.37}$$

$$\vartheta_{n,m}^{(1)} = x_1(\tilde{t}) - nq_1 \ , \quad \tilde{t} \in \left\{ t \mid |x_1(t) - nq_1| \leq \frac{q_1}{2}, |x_2(t) - mq_2| \leq \frac{q_2}{2} \right\} \tag{3.38}$$

$$\vartheta_{m,n}^{(2)} = x_2(\tilde{t}) - mq_2 \ , \quad \tilde{t} \in \left\{ t \mid |x_2(t) - mq_2| \leq \frac{q_2}{2}, |x_1(t) - nq_1| \leq \frac{q_1}{2} \right\} . \tag{3.39}$$

Here, $\vartheta_n^{(1)}$ and $\vartheta_m^{(2)}$ are the discretization errors that resulted in a discrete value of $\bar{x}_1 = nq$ and $\bar{x}_2 = mq$ accordingly, where n and m are integers. Consequently, $\vartheta_{n,m}^{(1)}$ and $\vartheta_{m,n}^{(2)}$ are discretization errors that led to a value of $\bar{x}_1 = nq$ and $\bar{x}_2 = mq$, while the other (correlated) time series was simultaneously discretized to $\bar{x}_2 = mq$ and $\bar{x}_1 = nq$. In all cases, $\tilde{t}$ is the set of time points at which these actual discretizations occur.

Using the interpolated distribution functions $\varrho_{x_1}(x(t))$ and $\varrho_{x_1}(y(t))$ and the interpolated joint distribution function $\varrho_{x_1,x_2}(x(t), y(t))$, the average discretization errors can be calculated as

$$\left\langle \vartheta_n^{(1)} \right\rangle = \frac{\int_{q_1(n-\frac{1}{2})}^{q_1(n+\frac{1}{2})} (z - nq_1)\varrho_{x_1}(z)\, dz}{\int_{q_1(n-\frac{1}{2})}^{q_1(n+\frac{1}{2})} \varrho_{x_1}(z)\, dz} \tag{3.40}$$

$$\left\langle \vartheta_m^{(2)} \right\rangle = \frac{\int_{q_2(m-\frac{1}{2})}^{q_2(m+\frac{1}{2})} (z - mq_2)\varrho_{x_2}(z)\, dz}{\int_{q_2(m-\frac{1}{2})}^{q_2(m+\frac{1}{2})} \varrho_{x_2}(z)\, dz} \tag{3.41}$$

$$\left\langle \vartheta_{n,m}^{(1)} \right\rangle = \frac{\int_{q_x(n-\frac{1}{2})}^{q_1(n+\frac{1}{2})} (z - nq_1)\varrho_{x_1,y_2}(z, mq_2)\, dz}{\int_{q_1(n-\frac{1}{2})}^{q_1(n+\frac{1}{2})} \varrho_{x_1,x_2}(z, mq_2)\, dz} \tag{3.42}$$

$$\left\langle \vartheta^{(2)}_{m,n} \right\rangle = \frac{\int_{q_2(m-\frac{1}{2})}^{q_2(m+\frac{1}{2})} (z - mq_2)\varrho_{x_1,x_2}(nq_1, z)\, dz}{\int_{q_2(m-\frac{1}{2})}^{q_2(m+\frac{1}{2})} \varrho_{x_1,x_2}(nq_1, z)\, dz} , \tag{3.43}$$

where

$$\int_{-\infty}^{+\infty} \varrho_{x_1,x_2}(x_1(t), z)\, dz = \varrho_{x_1}(x_1(t)) \quad \text{and} \tag{3.44}$$

$$\int_{-\infty}^{+\infty} \varrho_{x_1,x_2}(z, x_2(t))\, dz = \varrho_{x_2}(x_2(t)) \; . \tag{3.45}$$

Therefore the overall average discretization errors can be written as

$$\left\langle \vartheta^{(1)} \right\rangle \approx \frac{1}{T} \sum_{n=N_-}^{N_+} T_n \left\langle \vartheta^{(1)}_n \right\rangle \tag{3.46}$$

$$\left\langle \vartheta^{(2)} \right\rangle \approx \frac{1}{T} \sum_{m=M_-}^{M_+} T_m \left\langle \vartheta^{(2)}_m \right\rangle , \tag{3.47}$$

where T_n and T_m are the number of values that have been discretized to nq_1 and mq_2. Here, $q_1 N_-$ represents the minimum of the discretized time series $\bar{x}_1(t)$. $q_1 N_+$ is its maximum (M_- and M_+ is analogously defined).

Now we can calculate the discretization terms of Eq. (3.33). We begin with:

$$\operatorname{cov}\left(\bar{x}_1, \vartheta^{(2)}\right) = \left\langle \bar{x}_1 \vartheta^{(2)} \right\rangle - \langle \bar{x}_1 \rangle \left\langle \vartheta^{(2)} \right\rangle \tag{3.48}$$

$$= \frac{1}{T} \sum_{n=N_-}^{N_+} \sum_{m=M_-}^{M_+} \sum_{\tilde{t}=0}^{T_{n,m}} \left(nq_1 \vartheta^{(2)}_m(\tilde{t}) \right) - \langle \bar{x}_1 \rangle \left\langle \vartheta^{(2)} \right\rangle \tag{3.49}$$

$$= \frac{q_1}{T} \sum_{n=N_-}^{N_+} n \sum_{m=M_-}^{M_+} T_{n,m} \left\langle \vartheta^{(2)}_{m,n} \right\rangle - \langle \bar{x}_1 \rangle \left\langle \vartheta^{(2)} \right\rangle \; . \tag{3.50}$$

T is the length of the whole time series, while $T_{n,m}$ is the number of synchronous pairs of both time series, which are discretized to nq_1 and mq_2. We index these pairs with $\tilde{t}$ referring to these certain points in time.

Analogously, the other discretization terms of Eq. (3.33) can be calculated as

$$\operatorname{cov}\left(\bar{x}_2, \vartheta^{(1)}\right) = \frac{q_2}{T} \sum_{m=M_-}^{M_+} m \sum_{n=N_-}^{N_+} T_{n,m} \left\langle \vartheta^{(1)}_{n,m} \right\rangle - \langle \bar{x}_2 \rangle \frac{1}{T} \sum_{n=N_-}^{N_+} T_n \left\langle \vartheta^{(x)}_n \right\rangle \tag{3.51}$$

$$\operatorname{cov}\left(\bar{x}_1, \vartheta^{(1)}\right) = \frac{q_1}{T} \sum_{n=N_-}^{N_+} T_n n \left\langle \vartheta^{(1)}_n \right\rangle - \langle \bar{x}_1 \rangle \frac{1}{T} \sum_{n=N_-}^{N_+} T_n \left\langle \vartheta^{(1)}_n \right\rangle \tag{3.52}$$

$$\operatorname{cov}\left(\bar{x}_2, \vartheta^{(2)}\right) = \frac{q_2}{T} \sum_{m=M_-}^{M_+} T_m m \left\langle \vartheta^{(2)}_m \right\rangle - \langle \bar{x}_2 \rangle \frac{1}{T} \sum_{m=M_-}^{M_+} T_m \left\langle \vartheta^{(2)}_m \right\rangle \tag{3.53}$$

$$\operatorname{var}\left(\vartheta^{(1)}\right) \approx \frac{q_1^2}{12} \tag{3.54}$$

$$\operatorname{var}\left(\vartheta^{(2)}\right) \approx \frac{q_2^2}{12} \,. \tag{3.55}$$

The terms (3.54) and (3.55) are estimated under the assumption that the discretization errors are uniformly distributed. Usually, the remaining term $\operatorname{cov}(\vartheta^{(1)}_n, \vartheta^{(2)}_m)$ cannot be calculated with the distribution functions as it contains the correlation between the discretization errors. This value is not necessarily connected to the correlation of the whole time series either. Yet, we will show in the next section, that this term is negligible in the present context.

Thus, we have shown that the error caused by the discretization can be estimated by decomposing the correlation coefficient and approximating the mean discretization errors. This is achieved by interpolating the discrete distributions.

3.2.4 Estimating Correlations of Discretized Price Changes

We now turn to the specific situation on the stock market. The situation differs, when applying the method from the previous section to stock price changes. On price changes, the discretization process does not take place on the actual observable. Instead the price change ΔS is a difference for two prices $S(t)$ and $S(t + \Delta t)$ that are discretized by the tick-size q.

Therefore, the discretization error on a specific price difference $\Delta S'$ can be in the range from $-q$ to q. However, the probability that a certain value is from a price difference within this range is not constant. It is described by a triangular-shaped distribution (See Fig. 3.9a). This is evident, as the distribution error is the difference of two uniformly distributed discretization errors. The normalized triangular distribution ϱ_{Tri} around a certain price change $\Delta S'$ vanishes at $\Delta S' - q$ and $\Delta S' + q$ and has the value $1/q$ at its maximum at $\Delta S'$. It reads

$$\varrho_{\text{Tri}}(x, \Delta S') = \begin{cases} \frac{x-\Delta S'+q}{q^2} & (\Delta S' - q) \leq x < \Delta S' \\ \frac{-x+\Delta S'+q}{q^2} & (\Delta S' + q) \geq x \geq \Delta S' \\ 0 & \text{else} . \end{cases} \tag{3.56}$$

The average discretization errors have now to be calculated with the product of the triangular distribution ϱ_{Tri} and the interpolated price change distributions $\varrho_{\Delta S^{(1)}}$, $\varrho_{\Delta S^{(2)}}$ (and proper normalization). Thus,

$$\left\langle \vartheta_n^{(1)} \right\rangle = \frac{\int_{q_1(n-1)}^{q_1(n+1)} (z - nq_1)\varrho_{\Delta S^{(1)}}(z)\varrho_{\text{Tri}}(z, nq_1)\, dz}{\int_{q_1(n-1)}^{q_1(n+1)} \varrho_{\Delta S^{(1)}}(z)\varrho_{\text{Tri}}(z, nq_1)\, dz} , \tag{3.57}$$

$$\left\langle \vartheta_{n,m}^{(1)} \right\rangle = \frac{\int_{q_1(n-1)}^{q_1(n+1)} (z - nq_1)\varrho_{\Delta S^{(1)},\Delta S^{(2)}}(z, mq_2)\varrho_{\text{Tri}}(z, nq_1)\, dz}{\int_{q_1(n-1)}^{q_1(n+1)} \varrho_{\Delta S^{(1)},\Delta S^{(2)}}(z, mq_2)\varrho_{\text{Tri}}(z, nq_1)\, dz} , \tag{3.58}$$

while $\left\langle \vartheta_m^{(2)} \right\rangle$ and $\left\langle \vartheta_{m,n}^{(2)} \right\rangle$ are analogously defined.

Fig. 3.9b shows exemplarily the product of a triangular distribution and a power-law distribution. The denominator in Eq. (3.57) refers to the area under this curve. The triangular distribution also needs to be included in the fitting process. Thus, the difference function becomes

$$\begin{aligned} f_{\Delta S}(\varrho_{\Delta S}, \varrho_{\Delta \bar{S}}) &= \sum_{n=N_-}^{N_+} \left[\int\limits_{q_S(n-1)}^{q_S(n+1)} \varrho_{\text{Tri}}(z, nq_S) \left[\varrho_{\Delta S}(z) - \varrho_{\Delta \bar{S}}(nq_S)\right] dz \right] \\ &= \sum_{n=N_-}^{N_+} \left[\int\limits_{q_S(n-1)}^{q_S(n+1)} \varrho_{\text{Tri}}(z, nq_S)\varrho_{\Delta S}(z)\, dz - \varrho_{\Delta \bar{S}}(nq_S) \right] . \end{aligned} \tag{3.59}$$

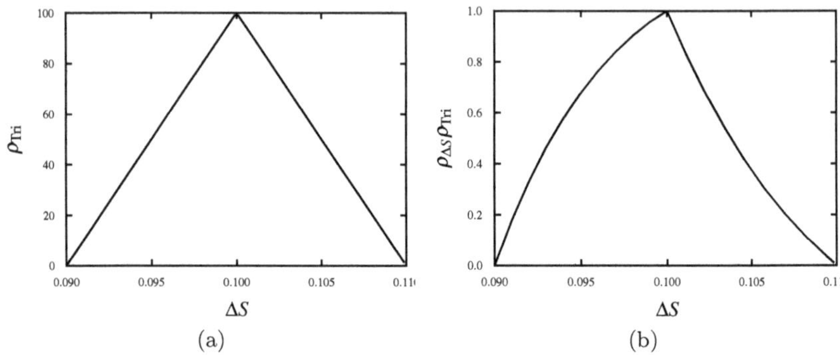

Figure 3.9: Exemplary distribution of discretization errors around a price change of $\Delta S = 0.1$ and a tick-size of $q_{\Delta S} = 0.01$ (a). Fig. (b) shows the product with power-law distribution given by $\rho_{\Delta S}(x) = 10x^{-6}$

Where $\varrho_{\Delta \bar{S}}$ refers to the discretized distribution. ϱ_{Tri} acts like a weighting function in the residual measure. It provides a weight corresponding to the probability that the difference of the originating discretization errors result in the value z.

Now, we are able to estimate the correlation discretization error with the previously defined Eqs. (3.50) to (3.55).

3.2.5 Estimating Correlations of Discretized Returns

When calculating the correlation of financial returns as defined in Eq. (3.21) the situation becomes more complex. Here, we also have to take the prices into account. The discretized correlation coefficient in Eq. (3.33) for two return time series $r^{(1)}$ and $r^{(2)}$ with underlying tick-sizes q_1 and q_2 reads

$$\begin{aligned}\widehat{\mathrm{corr}}_{\mathrm{tick}}(r^{(1)}, r^{(2)}) =& \frac{1}{\hat{\sigma}^{(1)}\hat{\sigma}^{(2)}} \left[\mathrm{cov}\left(\bar{r}^{(1)}, \bar{r}^{(2)}\right) + \mathrm{cov}\left(\frac{\Delta \bar{S}^{(1)}}{S^{(1)}}, \frac{\vartheta^{(2)}}{S^{(2)}}\right)\right.\\ &\left. + \mathrm{cov}\left(\frac{\Delta \bar{S}^{(2)}}{S^{(2)}}, \frac{\vartheta^{(1)}}{S^{(1)}}\right) + \mathrm{cov}\left(\frac{\vartheta^{(1)}}{S^{(1)}}, \frac{\vartheta^{(2)}}{S^{(2)}}\right)\right] \end{aligned} \tag{3.60}$$

with

$$\hat{\sigma}^{(i)} = \sqrt{\mathrm{var}\left(\bar{r}^{(i)}\right) + \mathrm{var}\left(\frac{\vartheta^{(i)}}{S^{(i)}}\right) + 2\mathrm{cov}\left(\frac{\Delta \bar{S}^{(i)}}{S^{(i)}}, \frac{\vartheta^{(i)}}{S^{(i)}}\right)}\,. \tag{3.61}$$

Here, $\bar{r}^{(1)}$ and $\bar{r}^{(2)}$ refer to the discretized return time series. Analogously to the correlation between price changes, the individual terms can be estimated, but in addition, the starting prices $S^{(1)}$ and $S^{(2)}$ need to be parameterized. We use the variables k and l for this. $q_1 K_-$ represents the minimum price within the observed time series, while $q_1 K_+$ represents the maximum price. $T_{n,m,k,l}$ represents the number of pairs whose returns equal $(q_1 n)/(q_1 k) = n/k$ and m/l. Similar to that, $T_{n,k}$ refers to the number of returns (from a single time series) that are equal to n/k. Thus, we obtain

$$\operatorname{cov}\left(\frac{\Delta\bar{S}^{(1)}}{S^{(1)}}, \frac{\vartheta^{(2)}}{S^{(2)}}\right) \approx \frac{q_1}{T} \sum_{n=N_-}^{N_+} n \sum_{m=M_-}^{M_+} q_2 \sum_{k=K_-}^{K_+} \sum_{l=L_-}^{L_+} T_{n,m,k,l} \frac{\left\langle \vartheta^{(2)}_{m,n} \right\rangle}{kl} - \left\langle \frac{\Delta\bar{S}^{(1)}}{S^{(1)}} \right\rangle \left\langle \frac{\vartheta^{(2)}}{S^{(2)}} \right\rangle \tag{3.62}$$

$$\operatorname{cov}\left(\frac{\Delta\bar{S}^{(2)}}{S^{(2)}}, \frac{\vartheta^{(1)}}{S^{(1)}}\right) \approx \frac{q_2}{T} \sum_{m=M_-}^{M_+} m \sum_{n=N_-}^{N_+} q_1 \sum_{k=K_-}^{K_+} \sum_{l=L_-}^{L_+} T_{n,m,k,l} \frac{\left\langle \vartheta^{(1)}_{n,m} \right\rangle}{kl} - \left\langle \frac{\Delta\bar{S}^{(2)}}{S^{(2)}} \right\rangle \left\langle \frac{\vartheta^{(1)}}{S^{(1)}} \right\rangle \tag{3.63}$$

$$\operatorname{cov}\left(\frac{\Delta\bar{S}^{(1)}}{S^{(1)}}, \frac{\vartheta^{(1)}}{S^{(1)}}\right) \approx \frac{q_1}{T} \sum_{n=N_-}^{N_+} \sum_{k=K_-}^{K_+} T_{n,k} \frac{n}{k^2} \left\langle \vartheta^{(1)}_{n} \right\rangle - \left\langle \frac{\Delta\bar{S}^{(1)}}{S^{(1)}} \right\rangle \left\langle \frac{\vartheta^{(1)}}{S^{(1)}} \right\rangle \tag{3.64}$$

$$\operatorname{cov}\left(\frac{\Delta\bar{S}^{(2)}}{S^{(2)}}, \frac{\vartheta^{(2)}}{S^{(2)}}\right) \approx \frac{q_2}{T} \sum_{n=N_-}^{N_+} \sum_{k=K_-}^{K_+} T_{n,k} \frac{n}{k^2} \left\langle \vartheta^{(2)}_{n} \right\rangle - \left\langle \frac{\Delta\bar{S}^{(2)}}{S^{(2)}} \right\rangle \left\langle \frac{\vartheta^{(2)}}{S^{(2)}} \right\rangle \tag{3.65}$$

and

$$\operatorname{var}\left(\frac{\vartheta^{(1)}}{S^{(1)}}\right) \approx \frac{q_1^2}{6} \left\langle \frac{1}{(S^{(1)})^2} \right\rangle \tag{3.66}$$

$$\operatorname{var}\left(\frac{\vartheta^{(2)}}{S^{(2)}}\right) \approx \frac{q_2^2}{6} \left\langle \frac{1}{(S^{(2)})^2} \right\rangle . \tag{3.67}$$

The terms $\langle \vartheta^{(1)}/S^{(1)} \rangle$ and analogously $\langle \vartheta^{(2)}/S^{(2)} \rangle$ in Eqs. (3.62) to (3.65) can be estimated as

$$\left\langle \frac{\vartheta^{(1)}}{S^{(1)}} \right\rangle \approx \frac{1}{T} \sum_{n=N_-}^{N_+} q_1 \sum_{k=K_-}^{K_+} T_{n,k} \frac{\left\langle \vartheta_n^{(1)} \right\rangle}{k} . \tag{3.68}$$

We note that the correlation between ΔS and S is neglected in this approximation. Also the discretization of the prices in the denominator of the return is not compensated. However, the model results in the next section demonstrate that this simplification only induces a minor error. This is indicated by the approximate-sign in Eqs. (3.62) to (3.68).

Also the impact of specific trading strategies can be calculated using the presented modeling. Here, the distortion of correlation coefficients, i.e., the distribution of discretization errors (Eq. (3.56)) needs to be chosen in a suitable manner.

3.3 Combined Compensation

Having presented compensation methods for distortions of correlations due to asynchronous time series and due to the tick-size, we now combine both findings. The compensation of asynchrony acts on each term of the Pearson correlation coefficient for every point in time. The tick-size compensation, in contrast, acts on the Pearson correlation coefficient as a whole in terms of the time series, but it acts on every occurring price change individually. Both effects superimpose, as illustrated in Fig. 3.10 on empirical data. The horizontal axis shows the product of normalized 1-min returns for each point in 2007 (overnight returns are excluded). The vertical axis shows the corresponding fractional overlap of each return pair. The discretization effects are visible in the center, superimposed with the asynchronous characteristics. Similar to the findings of Szpiro [117] for single stocks, the tick-size induces a nanostructure on the terms of the Person correlation coefficient for two stocks.

The simultaneous compensation of both effects can be achieved by com-

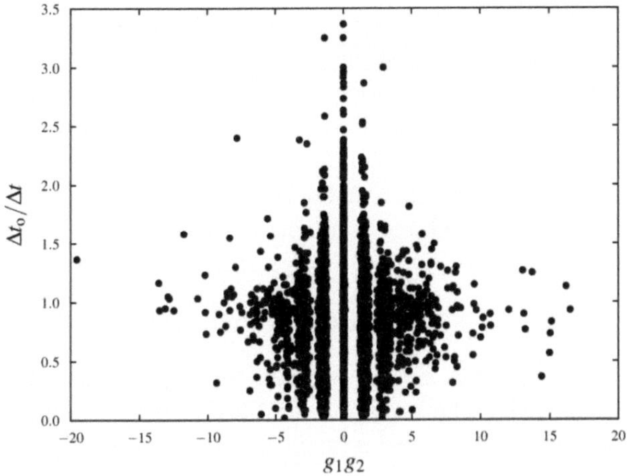

Figure 3.10: Product of normalized return pairs versus their fractional overlap for 1-min returns of the shares of Novell Inc. and Unisys Corp.in 2007. The average fractional overlap is 0.76.

bining both presented compensations. It reads

$$\begin{aligned}\widehat{\mathrm{corr}}(r^{(1)}, r^{(2)}) = \frac{1}{\hat{\sigma}^{(1)}_{\Delta t}\hat{\sigma}^{(2)}_{\Delta t}} &\left\langle \bar{r}^{(1)}_{\Delta t}\bar{r}^{(2)}_{\Delta t}\frac{\Delta t}{\Delta t_{\mathrm{o}}}\right\rangle \left\langle \frac{\Delta t}{\Delta t_{\mathrm{o}}}\right\rangle \\ &\times \left[\mathrm{cov}\left(\frac{\Delta\bar{S}^{(1)}_{\Delta t}}{S^{(1)}_{\Delta t}}, \frac{\vartheta^{(2)}}{S^{(2)}_{\Delta t}}\right)\right. \\ &+ \mathrm{cov}\left(\frac{\Delta\bar{S}^{(2)}_{\Delta t}}{S^{(2)}_{\Delta t}}, \frac{\vartheta^{(1)}}{S^{(1)}_{\Delta t}}\right) \mathrm{cov}\left(\frac{\vartheta^{(1)}}{S^{(1)}_{\Delta t}}, \frac{\vartheta^{(2)}}{S^{(2)}_{\Delta t}}\right) \\ &\left. - \left\langle \bar{r}^{(1)}_{\Delta t}\right\rangle \left\langle \bar{r}^{(2)}_{\Delta t}\right\rangle \right], \end{aligned} \tag{3.69}$$

which is the final result.

3.4 Model Results

Before applying the method to empirical data, we study it in a model setup. Subsequently, we apply the compensation methods to empirical data from the NYSE's TAQ database [43] to estimate the impact of the presented causes on the distortion correlations.

While first order autocorrelations are in this context insignificantly small [118], second order autocorrelations or "volatility clustering" represent a strong characteristic of return time series and led to the development of autoregressive models, such as GARCH [119, 120]. Thus, we start by generating an underlying correlated time series using a GARCH(1,1) model, as introduced in [69],

$$r^{(i)}(t) = \sigma^{(i)}(t)\left(\sqrt{c}\,\eta(t) + \sqrt{1-c}\,\varepsilon_i(t)\right) . \tag{3.70}$$

Here $r^{(i)}(t)$ stands for the return of the i-th stock at time t and c is the correlation coefficient. The random variables η and $\varepsilon^{(i)}$ are drawn from a Gaussian distribution. $\eta(t)$ is identical for all stocks; it induces the correlation. The ε_i are individual for each stock. $\sigma^{(i)}(t)$ is the non-constant variance, given by a GARCH(1,1) process,

$$\left(\sigma^{(i)}(t)\right)^2 = \alpha_0 + \alpha_1\left(r^{(i)}(t-1)\right)^2 + \beta_1\left(\sigma^{(i)}(t-1)\right)^2 . \tag{3.71}$$

The initial parameters of the GARCH(1,1) process are chosen as $\alpha_0 = 2.4 \times 10^{-4}$, $\alpha_1 = 0.15$ and $\beta_1 = 0.84$. If $\sigma^{(i)}$ is set to unity, Eq. (3.70) becomes Noh's realization [76], likely familiar to the physics community, of the Capital Asset Pricing Model (CAPM) [75].

Two return time series $r^{(1)}$ and $r^{(2)}$ are generated representing two correlated stocks. The total lengths of these time series is chosen as 7.2×10^6, corresponding to a return interval Δt of one second during one trading year. From these returns, we generate two underlying price time series $\tilde{S}^{(1)}$ and $\tilde{S}^{(2)}$. We set the starting prices at $t = 0$ to 1000. c is chosen as 0.4.

To model the asynchronous trade processes, these prices are sampled independently using exponentially distributed waiting times with average values typical for the stock market. To model the asynchrony, we choose the average waiting times as 15 and 25 data points (equivalent to seconds in this setup).

Eventually, we construct the time series of returns from these prices using return intervals from 60 data points (corresponding to 1 minute) to 1800 data points (corresponding to 30 minutes).

When calculating the correlation of returns of the sampled time series $\tilde{S}$ using different return intervals Δt, the correlation scales down. This behavior is very similar to the Epps effect in empirical data. It occurs only because of the asynchrony of the trading times.

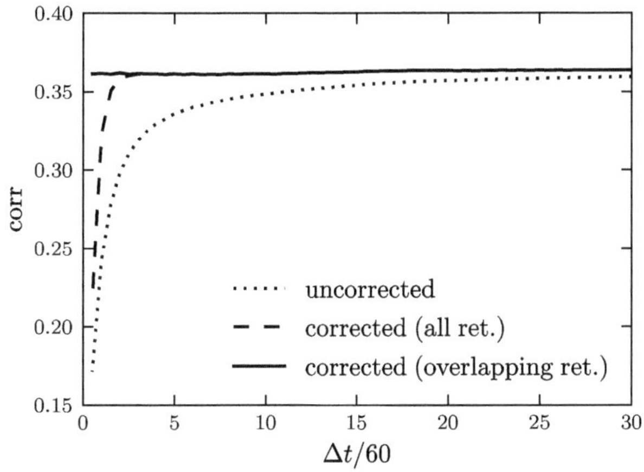

Figure 3.11: Compensation of asynchrony effects within the model. The dashed line represents the correlation coefficient that is corrected by the overlap. The solid line regards in addition only returns, in which time intervals trades occurred.

To model the discretization, we round the prices to integer values. An integer price of, for example, 1000 then corresponds to a price of 10.00 with a tick-size of 0.01. The thus obtained time series features both, asynchrony and discretization. When including the discretization by the tick-size, the downscaling is even more emphasized. When considering synchronous non-discretized time series, we set the waiting times to the same value for each time series (e.g. every 60 seconds). To demonstrate the developed compensation techniques, we first evaluate them separately, using model setups that only exhibit the relevant feature. This is performed in sections 3.4.1 and 3.4.2. The next step is performed in section 3.4.3 where we combine both effects simultaneously in a model that features asynchrony as well as discretization effects.

3.4.1 Asynchrony Compensation

First, we consider a model for the impact of asynchronous correlations while neglecting discretization effect. The model results are shown in Fig. 3.11. The black line corresponds to the asynchrony's impact on the Epps effect in our model. Using the developed asynchrony compensation, we are ably to compensate the decline of the correlation within the model, as illustrated by the dashed line.

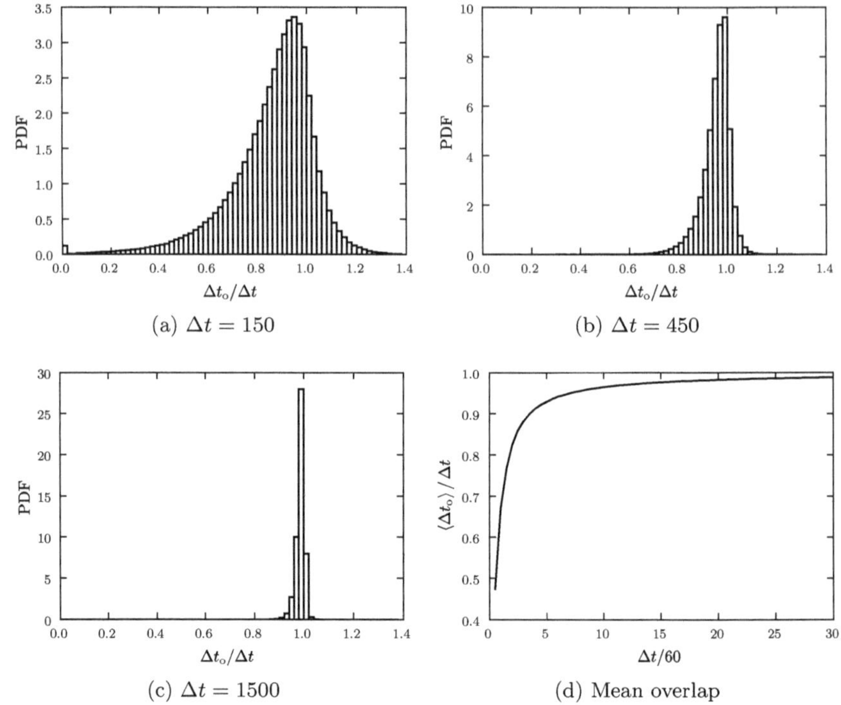

Figure 3.12: Distribution of the overlaps for different return intervals $\Delta t = 150$ (a), 450 (b) and 1500 (c) data points (corresponding to 2.5, 7.5 and 25 minutes) in the model setup. Towards larger return intervals, the distribution sharpens, as well as its mean converges to one (d).

However, there is a remaining effect that still causes a downscaling of correlations for very small return intervals. This behavior occurs when the price of either of the stocks did not change during the return interval and therefore the corresponding return equals zero. Evidently, this event becomes more probable on smaller return intervals Δt. This remaining downscaling coincides with the cumulative estimator described by Hayashi and Yoshida [97]. It can also be expressed in the formalism used here. It reads

$$\mathrm{corr}(r^{(1)}_{\Delta t}, r^{(2)}_{\Delta t})_{\mathrm{HY}} = \mathrm{corr}(r^{(1)}_{\Delta t}, r^{(2)}_{\Delta t})\Big|_{(\gamma_1^{(1)}(t) \neq \gamma^{(1)}(t+\Delta t)) \wedge (\gamma^{(2)}(t) \neq \gamma^{(2)}(t+\Delta t))} . \tag{3.72}$$

Therefore, when combining both estimators, and thus only regarding returns with overlapping time intervals, the remaining scaling behavior for very small returns can be compensated as well as indicated by the solid line in Fig. 3.11.

However, the main contribution is due to the asynchrony and can be compensated by calculating the overlap. An insight on the component of the compensation, i.e., the fractional overlaps, for each term of the correlation coefficient, is given in Fig. 3.12. The histograms demonstrate, that the overlap can also be larger than the actual return interval, implying that terms with such overlaps are corrected downwards. However, on average, the fractional overlap is smaller than the return interval, as shown in Fig. 3.12d. Therefore, the compensation can amplify a specific term of the correlation coefficient as well as it can attenuate it. The impact of Hayashi and Yoshida's estimator is bound to the amount of null returns. This corresponds to the small peak at $\Delta t_\mathrm{o}/\Delta t = 0$ in the histogram shown in Fig. 3.12a.

3.4.2 Tick-Size Compensation

Now we turn to the compensation of discretization effects. We set up our model to generate completely synchronous returns that are discretized.

As we know the actual discretization errors in the model setup precisely, we can use it to evaluate error estimates that we discussed in section 3.2.4. A comparison of the estimated average discretization errors with the actual discretization errors is shown in Fig. 3.13. The estimated values show an excellent agreement with the original values. The interpolation is restricted to a single Gaussian fit, as we know the type of the price change distributions in this case. Thus, we can verify the scope of the estimation itself, not the suitability of the interpolation.

Before we perform the compensation, we want to test how much impact each correction term (Eqs. (3.62) to (3.67)) has. We quantify the impact by calculating Eq. (3.60) and subtract the value of this expression with the regarded term set to zero. By this method we can see how the correlation coefficient changes, if a certain term of the discretization compensation is neglected (set to zero). This gives the opportunity to save computation time,

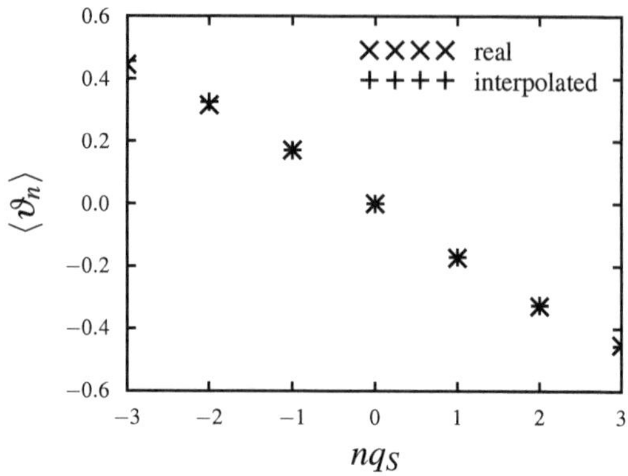

Figure 3.13: Benchmark of the error estimation: Comparison between actual and estimated discretization errors within the model setup.

as for large portfolios, the interpolation of the price change distribution is computationally very intensive.

Fig. 3.14 illustrates the results of this analysis for different start prices and correlation coefficients. A visual inspection of the subfigures reveals that only Eqs. (3.64) to (3.67) provide a sizable contribution to the compensation. Therefore, the compensation can be restricted to the calculation of these terms. This implies that the distortion of the correlation coefficient is mainly caused by an improper normalization of the returns, as the significant terms only appear in the correction of the standard deviations of each return.

Hence, the main contribution to the distortion of correlation coefficients in small return intervals is the overestimation of σ. Thus, we have

$$\widehat{\mathrm{corr}}_{\mathrm{tick}}(r_{\Delta t}^{(1)}, r_{\Delta t}^{(2)}) \approx \frac{\left\langle \bar{r}_{\Delta t}^{(1)} \bar{r}_{\Delta t}^{(2)} \right\rangle}{\hat{\sigma}_{\Delta t}^{(1)} \hat{\sigma}_{\Delta t}^{(2)}} \ . \tag{3.73}$$

Fig. 3.15 shows this overestimated σ and the tick-size-corrected $\hat{\sigma}$ versus the return interval Δt. This is consistent with the findings of Hansen and Lunde [121]. They demonstrate that the realized variance is overestimated on small return intervals due to microstructure noise.

Due to the convex shape of the price change distribution, the discretization errors are not distributed symmetrically. This effect grows with the impact of the discretization, i.e., smaller return intervals. Thus, the estima-

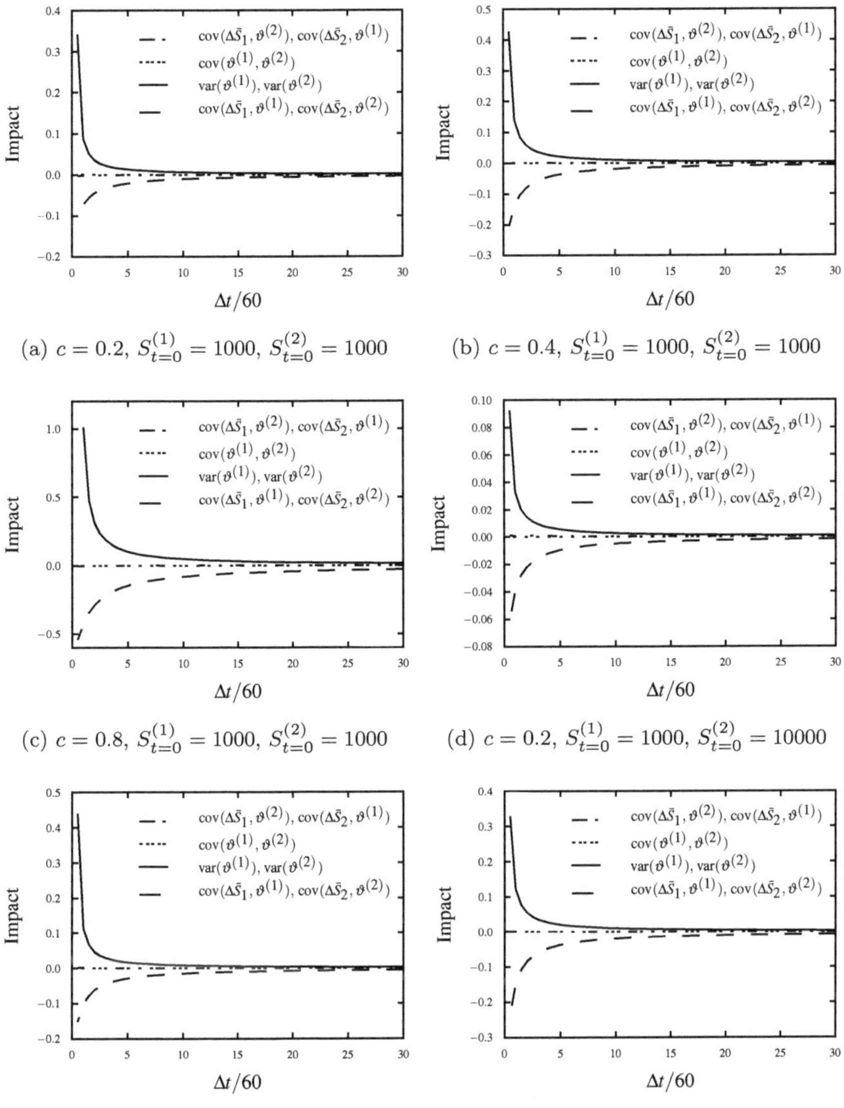

(a) $c = 0.2$, $S^{(1)}_{t=0} = 1000$, $S^{(2)}_{t=0} = 1000$

(b) $c = 0.4$, $S^{(1)}_{t=0} = 1000$, $S^{(2)}_{t=0} = 1000$

(c) $c = 0.8$, $S^{(1)}_{t=0} = 1000$, $S^{(2)}_{t=0} = 1000$

(d) $c = 0.2$, $S^{(1)}_{t=0} = 1000$, $S^{(2)}_{t=0} = 10000$

(e) $c = 0.4$, $S^{(1)}_{t=0} = 1000$, $S^{(2)}_{t=0} = 10000$

(f) $c = 0.8$, $S^{(1)}_{t=0} = 1000$, $S^{(2)}_{t=0} = 10000$

Figure 3.14: Impact of each term of the compensation method for the correlation coefficient between price changes. Evaluated using Noh's model, i.e., Eq. (3.70) with $\sigma^{(i)}(t) = 1$.

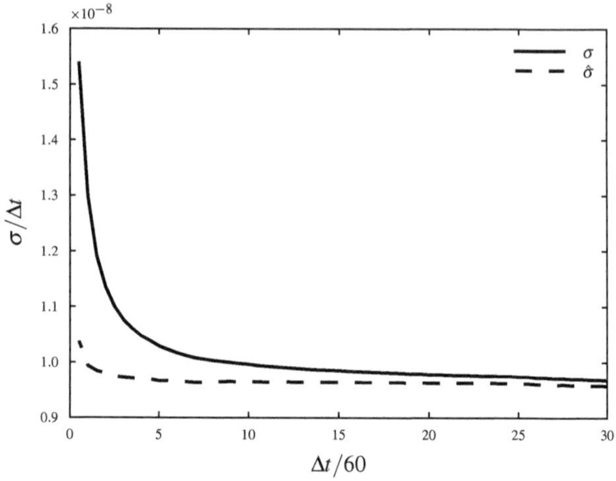

Figure 3.15: Overestimated standard deviation σ and tick-size-corrected standard deviation $\hat{\sigma}$ versus the return interval Δt in the model setup.

tion of variances on the discretized values is biased. This gives the largest contribution to the distortion of correlation coefficients due to discretized data. We can correct this behavior with the discussed tick-size compensation.

The model results are shown in Fig. 3.16. We are able to compensate the discretization effects. We first focus on the correlations between price changes. As shown in Fig. 3.16 (a) and (c), the correlation coefficient decays towards smaller price change intervals. Our model demonstrates that this effect can also contribute to the Epps effect. This effect becomes especially relevant when the ratio of the price to the tick-size is sufficiently small. It is remarkable that this scaling behavior is observed even though the time series are synchronous. The effect vanishes in our simulation, when both prices start with a value of 10 000, as Fig. 3.16e illustrates.

When applying the compensation method to return time series as illustrated in Figs. 3.16 (b,d,f), we are also able to correct the discretization error almost completely. The slight decay of the corrected correlation coefficient on very small return intervals is due to the discussed approximation. These are the negligence of the correlation between price changes and prices. In addition, even though the discretization of price changes is corrected, the

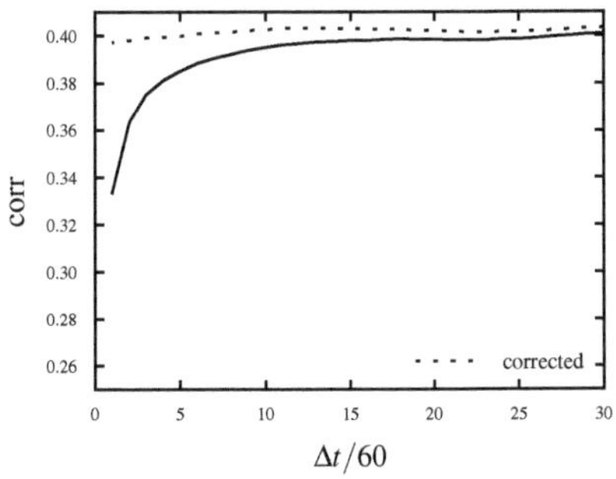

(a) $S^{(1)}_{t=0} = 1000$, $S^{(2)}_{t=0} = 1000$, $c = 0.4$

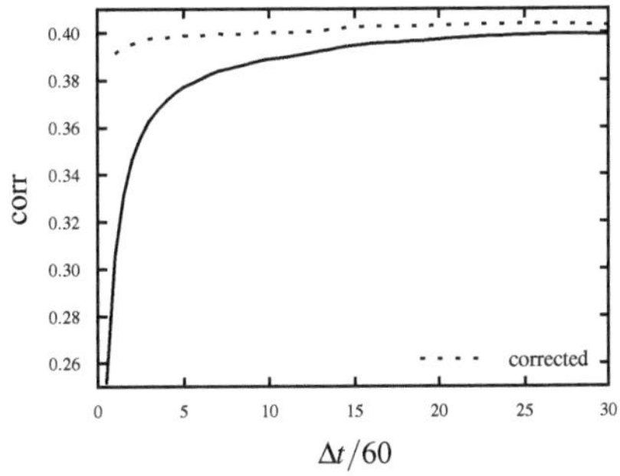

(b) $S^{(1)}_{t=0} = 1000$, $S^{(2)}_{t=0} = 1000$, $c = 0.4$

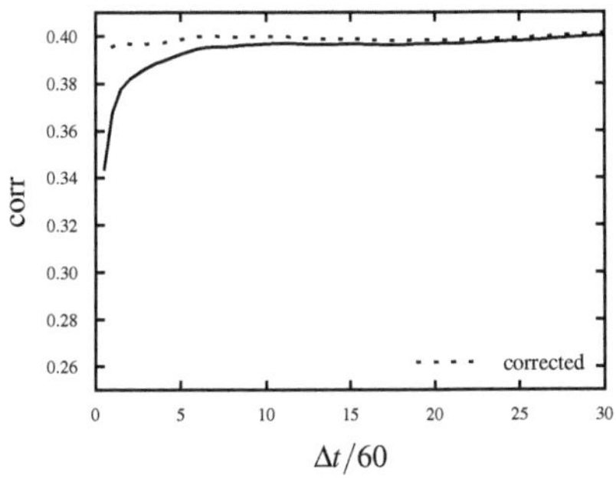

(c) $S^{(1)}_{t=0} = 1000$, $S^{(2)}_{t=0} = 10000$, $c = 0.4$

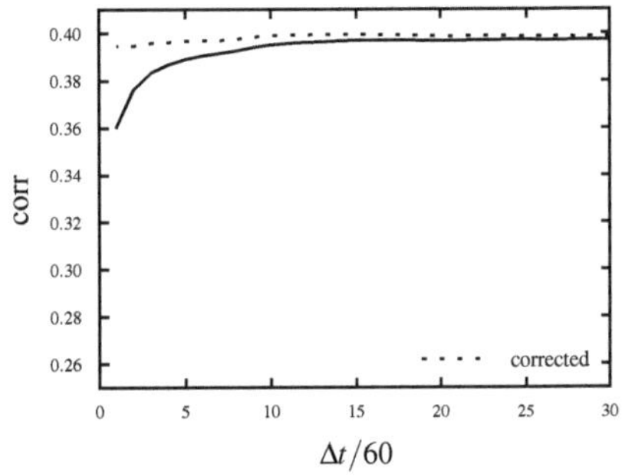

(d) $S^{(1)}_{t=0} = 1000$, $S^{(2)}_{t=0} = 10000$, $c = 0.4$

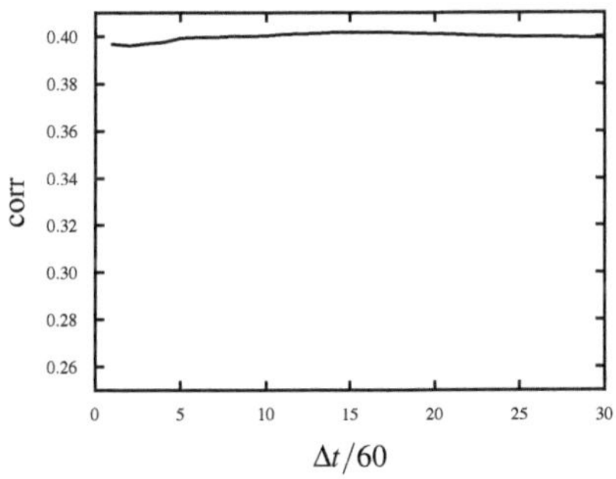

(e) $S^{(1)}_{t=0} = 10000$, $S^{(2)}_{t=0} = 10000$, $c = 0.4$

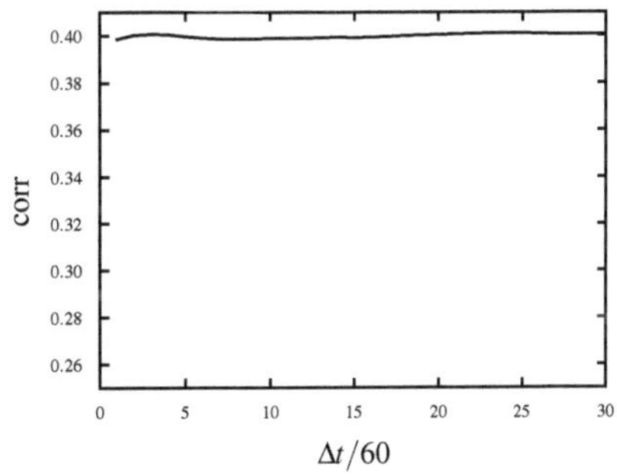

(f) $S^{(1)}_{t=0} = 10000$, $S^{(2)}_{t=0} = 10000$, $c = 0.4$

Figure 3.16: Scaling behavior of the correlation coefficient of price changes (a,c,e) and returns (b,d,f) due to the discretization error in the model setup. The dashed line represents the presented correction. Evaluated using Noh's model, i.e., Eq. (3.70) with $\sigma^{(i)}(t) = 1$.

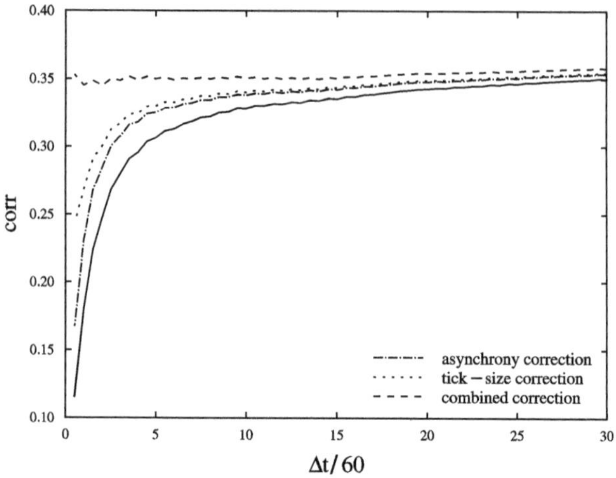

Figure 3.17: GARCH(1,1) Model results of compensation methods for the Epps effect.

price discretization in the denominator of the return is neglected. A further improvement of the compensation could be achieved by including these effects. However this would require further assumptions on the price process and would increase the necessary computing time dramatically. Thus, the presented compensation is restricted to this approximation, only making very few assumptions.

3.4.3 Combined Compensation

In the previous sections, we demonstrate the compensation of asynchrony and discretization effect separately. The tick-size distortion is mainly caused by an overestimation of the standard deviations. Thus, we can approximate the combined compensation given in Eq. (3.69) by

$$\widehat{\mathrm{corr}}(r_{\Delta t}^{(1)}, r_{\Delta t}^{(2)}) \approx \frac{\left\langle \bar{r}_{\Delta t}^{(1)} \bar{r}_{\Delta t}^{(2)} \frac{\Delta t}{\Delta t_{\circ}} \right\rangle}{\hat{\sigma}_{\Delta t}^{(1)} \hat{\sigma}_{\Delta t}^{(2)}} . \tag{3.74}$$

Fig. 3.17 illustrates the Epps effect in a model setup that exhibit both effects, i.e. asynchrony and discretization. Using the combined compensation of Eq. (3.69) we are able to compensate for both effects simultaneously which is confirmed by the dashed line in Fig. 3.17. The dotted and dash-dotted line represents the separate compensation of both effects. Separately, neither the asynchrony nor the tick-size compensation appear to have a profound impact at first sight. However, combining both methods allows a full compensation of the Epps effect in our model.

3.5 Empirical Results

After developing methods for compensating statistical errors in the calculation of correlations and verifying these methods in a model setup, now their application to empirical data is considered.

Similar to the previous section, we first turn to the asynchrony compensation in section 3.5.1. The results of the tick-size compensation are discussed in section 3.5.2. Eventually, we perform the combined compensation in section 3.5.3.

3.5.1 Asynchrony Compensation

It is difficult to isolate the Epps Effect on single stock pairs, as it can superimpose with other effects leading to other characteristics of the correlation coefficient than expected according to the Epps effect.

Hence, we classify two ensembles of stock pairs. After compensating the asynchrony effect for each pair, we build the average for the ensemble. We also plot the error bars representing twice the standard deviation 2σ of the compensation. By this method, we can show the scope of the asynchrony model and identify regions, in which other effects dominate. Furthermore, we deliberately chose the year 2007 for our empirical analysis because this year was relatively "quiet", with less fluctuation and extraordinary features than the forthcoming crisis-dominated years. With this approach, we can be sure that the results can be applied to a "typical" and do not rely on properties that only exist during financial crises. All data is extracted from the NYSE's TAQ database [43].

The first ensemble consists of stock pairs which provide the most stable correlation. Thereby we want to suppress those effects which are caused by a change in the correlation during the period in which the correlation coefficient is calculated. This ensemble represents ideal test conditions for the asynchrony compensation. To identify those stock pairs with a stable correlation, we calculate the correlation coefficient of 30 daily returns. After shifting this window in 1-day intervals through the year, we calculate the variance of the obtained correlation coefficients (var_{corr}). Then we identify the five stocks providing the smallest variance for each *Global Industry Classification System* (GICS) branch of the *Standard & Poor's* (S&P) 500 index. This results in an ensemble of 50 stocks as shown in appendix A.1.

As the correlation structure of stocks can be highly non-stationary, we also evaluate the asynchrony compensation without the restriction to stable correlations. For this purpose, we select a second ensemble consisting of 5 stock pairs of each GICS branch of the S&P 500 index, whose daily returns are providing the strongest correlation during the year 2007. These stocks include highly non-stationary correlations as indicated in appendix A.2.

Fig. 3.18 shows the ensemble average of the correlation coefficient and the asynchrony-compensated correlation coefficient for both ensembles in 2007 (250 trading days). Before averaging, the correlation coefficients for each stock have been normalized to the value at a return interval of $\Delta t = 40$ minutes.

When considering the whole ensemble we discover that the asynchrony has a pronounced impact on the Epps effect. The asynchrony effect seems to be the dominating cause for the Epps effect on return intervals down to approximately 10 minutes, where the remaining Epps effect is on average less than 3% of the correlation coefficient's saturation value at large return intervals. For smaller return intervals, other effects dominate, e.g. a lag between the time series of two stocks, as a recent study indicates [100].

However, the ensemble consists also of stocks which are very frequently traded, providing a very short average waiting time which results in a fractional overlap $\Delta_{t_o}(t)/\Delta t$ close to unity. Evidently the presented compensation only has a small impact on the correlation estimation of these stocks, as they are so frequently traded that their time series can almost be described as continuous. Naturally the presented compensation has the largest impact on less frequently traded stocks, as they actually show an asynchronous behavior. Many of the stocks in the S&P 500 index are being traded frequently. Thus, we would not expect a large impact of the asynchrony on "typical" S&P 500 stocks.

Within the statistical ensemble stock pairs can be found that either do not show an Epps effect or that are so infrequently traded that the assumption of an underlying timeline seems to be unreasonable. Even though the assumption of an underlying time series is a common and intuitive approach on this topic, it may not be valid for very infrequently traded stocks.

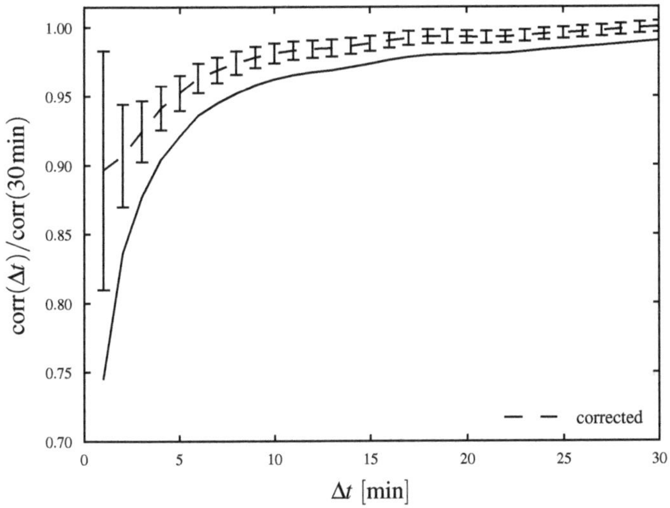

(a) Ensemble of most stable correlations

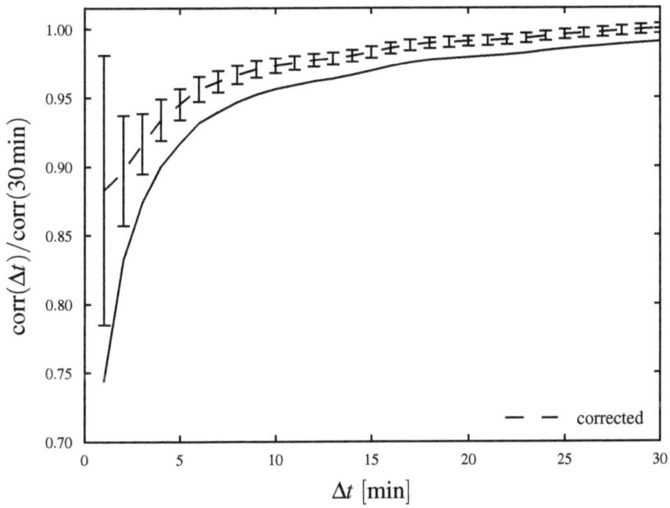

(b) Ensemble of highest correlations

Figure 3.18: Asynchrony-compensated correlations of two different stock ensembles. The data is normalized to its saturation value at 40 minutes. The error bars represents twice the standard deviation 2σ of the compensation.

When considering single stock pairs, we find that the asynchrony-compensation works well, if a distinguished Epps effect is found. In case of adopting the presented method as a black box model without looking at the scaling behavior of the correlation coefficient, we believe that a return interval of 5 minutes represents a good lower bound for the scope of this method.

3.5.2 Tick-Size Compensation

We now demonstrate that the price discretization can result in a sizable contribution to the Epps effect as well. As the mean price change per return interval decreases with the length of the interval (see, e.g., [122]), the width of the price change distribution decreases as well. While the tick-size remains constant, the discretization error increases. Hence, the tick-size should also have an impact on the Epps effect - especially for stocks that are traded at low prices.

But how large is the contribution of the discretization effect to the Epps effect? To answer this question, we apply the compensation to empirical data from the NYSE TAQ database [43]. Here, we use a power-law approach for the interpolation of the price change distribution, as the model results indicate that the discretization effects are mainly relevant for small return intervals. On small return intervals, power-law tails can describe the distribution satisfactory [20]. We perform a least squares fit of a and b in $\varrho_{\Delta S} = ax^{-|b|}$ for each value of the (discrete) distribution and their next two left and right neighbors individually. For the very central part of the distribution (consisting of three price changes) a Gaussian fit was performed.

It is particularly important that stock splits must not be corrected in order to maintain the correct tick-size. Of course, therefore overnight returns have to be excluded. To analyze the impact of the discretization effect, we construct two ensembles (see appendix A.4 and A.5) of stocks from the S&P 500 index. The first ensemble consists of stocks that are averagely priced between USD 0.01 and USD 10.00. The second ensemble consists of stocks that are on average priced between USD 10.01 and USD 20.00. Both ensembles are composed of 25 stock pairs providing the highest correlation during the year 2007 (based on daily data).

As Fig. 3.19 demonstrates, we are able to compensate the impact of the tick-size on the correlation coefficient in empirical market data. Certainly, the decay cannot be corrected completely with the presented method, as the discretization effect superimposes with other causes of the Epps effect such

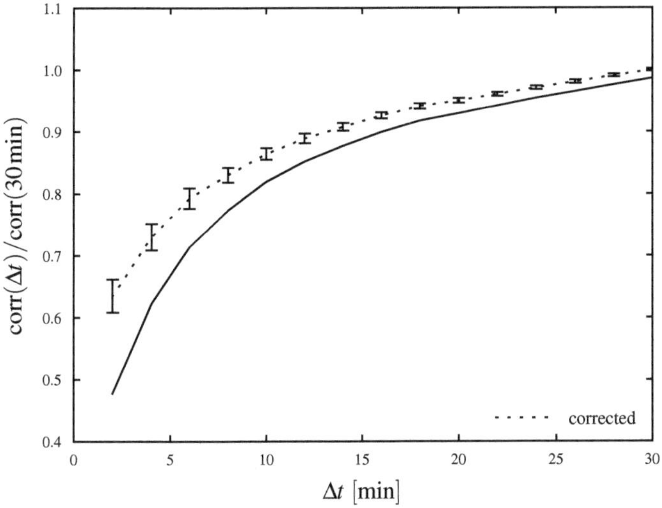

(a) Ensemble: USD 0.01–USD 10.00

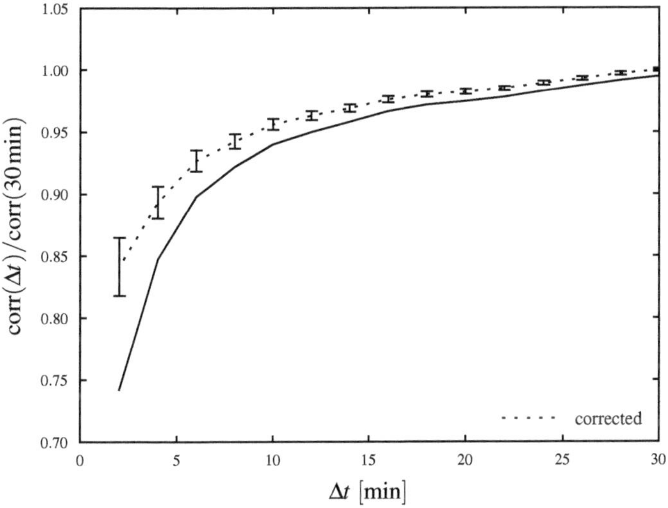

(b) Ensemble: USD 10.01–USD 20.00

Figure 3.19: Tick-size compensation of the correlation coefficient between two ensembles consisting of the 25 highest correlated stocks from the S&P 500 index that are averagely quoted within the region of USD 0.01–USD 10.00 and USD 10.01–USD 20.00, respectively. The correlation coefficients have been normalized to its saturation value at approximately 30 min. The error bars represents twice the standard deviation 2σ of the compensation.

as asynchronous or lagged time series. However, we were able to quantify the contribution of this particular effect to the Epps effect. Our results show, that the discretization effect can be responsible for up to 40% of the Epps effect, which we define as the difference between the correlation coefficient at a given time and its saturation value. The contribution is particularly large for stocks that are traded at low prices.

3.5.3 Combined Compensation

Eventually, we apply the combined compensation discussed in section 3.3 to the same ensembles as for the tick-size compensation in section 3.5.2. Results are shown in Fig. 3.20. The empirical results indicate that the two identified statistical effects can have a profound contribution to the Epps effect. For the first ensemble, a large portion of the Epps effect can be compensated. However there is a remaining downscaling effect that cannot be compensated with the presented techniques.

This is due to the fact that we choose our ensemble to exhibit large exposure to discretization. However, stocks that are traded on such low prices are sometimes being traded very infrequently that the assumption of an underlying time series is not justified and we thus are beyond the scope of the asynchrony compensation method. However, we can still compensate a large portion of the Epps effect and thus increase the precision of correlation estimation.

The second ensemble represents a more adequate candidate to isolate both, asynchrony and discretization effects. It consists of frequently traded stocks and is also exposed to a considerable amount of discretization. Here, a major portion of the Epps effect can be compensated. However, there is a remaining downscaling of the correlation coefficient towards return intervals smaller than 10 minutes.

Hence, this does not represent a full description of the Epps effect. However, both causes together can contribute to a significantly large portion of the Epps effect. In this case, we are able to compensate the Epps effect almost completely, only making very few and reasonable assumptions.

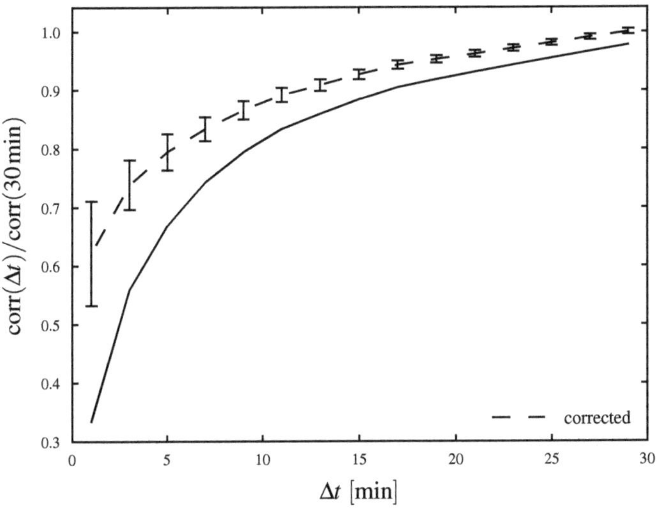

(a) Ensemble: USD 0.01–USD 10.00

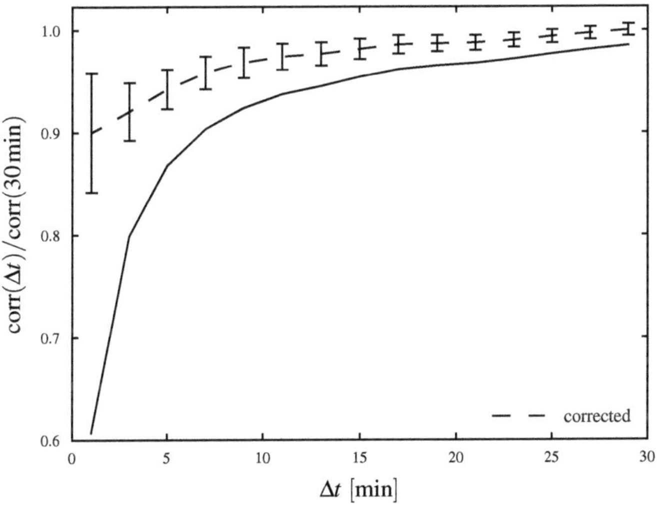

(b) Ensemble: USD 10.01–USD 20.00

Figure 3.20: Empirical results of combined compensation method for the Epps effect. Average over the 25 highest correlated stock pairs that were traded between USD 0.01 and USD 10.00 (a) and USD 10.01 and USD 20.00 (b) in 2007. The correlation coefficients have been individually normalized to the corrected value at $\Delta t = 30$ min. The error bars represents twice the standard deviation 2σ of the compensation.

3.6 Summary

We identified two purely statistical causes of the Epps effect. The asynchrony of time series as well as the tick-size have a major impact on the Epps effect. We developed two methods to compensate for these causes.

After evaluating the compensation methods in a model setup, we applied them to empirical market data under the assumption of an underlying time series with non-lagged correlations. The results clearly demonstrate that the asynchrony as well as the tick-size can have a huge impact on the Epps effect. They can even be the dominant cause.

However, this is not a full description of the Epps effect as there are certainly many phenomena contributing to it. In certain scenarios, other statistical properties of the time series or other causes for the Epps effect might dominate. The size of the error bars in Fig. 3.20 indicates that the asynchrony compensation does not give reliable results for return intervals below 3 minutes. Especially for very small return intervals, a lag between the time series of two stocks might be the dominating cause, as suggested by Tóth and Kertész [100]. There are even other, unknown mechanisms that lead to an inverse Epps effect, i.e., in some cases the correlation increases significantly for small return intervals. As these mechanism is not understood, an application in portfolio optimization, as discussed in section 2.1.3, on short return intervals might be difficult. This is due to the fact that in portfolio optimization, an overestimation (caused by an "inverse" Epps effect) of correlations, even only for a few stocks, is much more critical than an underestimation of correlations. Thus, before applying the compensation methods in portfolio optimization as black box model in terms of portfolio optimization, other market microstructure effects need to be understood.

For stocks that are infrequently traded at very low prices (often referred to as *penny-stocks*) the assumption of uniformly distributed discretization errors needs to be carefully reflected. It is possible that certain trading strategies dominate for those stocks leading to an asymmetrical distribution of discretization errors.

Nonetheless, the presented compensations significantly improve the estimation of financial correlations. These methods do not require parameter adjustments or model calibrations. Our empirical study indicates that the identified causes can contribute up to 75% of the Epps effect for stocks that are traded at low prices.

4 Credit Risk

In financial context, the majority of studies in econophysics are dealing with market risk, since many concepts in statistical physics are directly applicable. A type of risk that is fundamentally different from the other types of financial risks discussed in the introduction is represented by *credit risk* [123–127]. Modeling credit risk, i.e., the risk that an obligor fails to make a promised payment, is much more complex. As discussed in the previous chapters, the risk of a financial asset, e.g., a stock, is often expressed by its standard deviation. Due to the nearly symmetric shape of the return distribution of, e.g., stocks, by this definition, large positive returns are considered just as risky as large negative returns. An investor who uses volatility as a risk measure acts *risk averse.* However, due to the asymmetric shape of a credit portfolio's loss distribution, the variance is not suitable as a risk measure in this context.

The primary challenge is to model the loss distribution. It specifies the probability of a certain loss that is caused by obligors that do not pay back their debt. The loss distribution can be described as skewed and leptokurtic [123]. Especially the estimation of its tails, the probability of large losses is of central interest.

The financial crisis of 2008–2009 clearly revealed that an improper estimation of credit risk can lead to drastic effects on the world's economy. The vast underestimation of risks embedded in credits for the subprime housing markets induced a chain reaction that propagated into the worldwide economy. An improved estimation of credit risk is therefore of vital interest.

As shown by numerical simulations in a previous study of Schäfer et. al. [31], the concept of diversification, i.e., the reduction of risk by increasing the portfolio size, cannot be transferred to credit risk. In a portfolio of credits, such as collateralized debt obligations (CDOs), the effect of diversification is severely limited by the presence of even weak correlations.

In this chapter, we discuss this matter analytically. We estimate the risk embedded in a credit portfolio under the assumption of random correlations with average correlation level zero. This can provide a lower bound for the estimation of the risk in credit portfolios.

This chapter is organized as follows[1]. In section 4.1 a brief introduction on the concept of credits is given. We discuss three types of credit risk models in section 4.2. In chapter 4.3, we develop the structural credit risk model step by step, which we demonstrate in an application in section 4.4. We conclude our findings in section 4.5.

4.1 Phenomenology

A credit is a formalized process of lending money. If a *creditor* lends money to an *obligor*, the credit is a written agreement that the obligor will pay back the money. The creditor's reason to lend money is that the obligor does not only have to pay back the amount he borrowed, but also a surplus, typically defined by an interest rate and a risk compensation. The obligor takes credits because he wants to raise his capital. A very common type of credit is a *bond*. Bonds are being sold by many kinds of institutions, such as banks, companies, states, cities, etc. A simple category is the *zero-coupon* bond. If an investor buys such bonds from a bank, for a specific amount of money, technically, he/she lends money to the bank. The bank acts as an obligor. At *maturity*, a certain point in the future, defined in the bond contract, the bank will pay back the creditor not only the amount borrowed, but also the surplus, which is the profit of the investor. The amount that is due at maturity is called the *face value* of the bond. More complex types of bonds include, e.g., the periodical payment of an interest rate, the coupon, until maturity is reached.

At first glance, this seems to be an easy way for an investor to make profits by lending credits, but there are also risks involved. For example, the obligor might not be able to pay back his debt at maturity because he went bankrupt or for other reasons. This scenario is referred to as a *default*. Even if a default is relatively rare, it can cause huge losses for creditors. This is especially true if correlations are present in the creditor's portfolio of credits, because losses can occur simultaneously. The *default probability* is a central key in the estimation of credit risk.

The traditional way of quantifying credit risk is to determine the obligor's credit worthiness. A wide range of rating systems exists that aim at classifying credit worthiness in a system of levels. Prominent examples are the rating agencies *Moody's* and *Standard & Poor*. Rating agencies periodically analyze obligors to provide recent information on their credit worthiness.

[1] For details see Ref. [7].

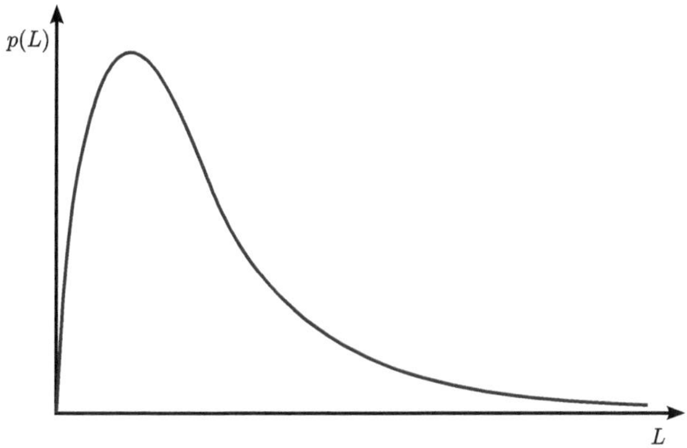

Figure 4.1: Sketch of a typical loss distribution $p(L)$. The distribution has a delta peak at $L = 0$ corresponding to the case of non-default.

It is evident, that these rating agencies have a huge impact on the worldwide economy, e.g., if they downgrade the credit rating of a whole country. Criticism arose during the financial crisis of 2008–2009, as the rating principles are not transparent and the agencies are not considered as neutral, i.e., they pursue interests of the country they are based in. Approaches are being made to estimate the *migration risk*, the probability of an obligor being upgraded or downgraded [128–130].

Often, a creditor does not only issue a single credit, but a large number of credits to various institutions and individuals. In this credit portfolio correlations represent an additional observable that influences the risk. For example, during a bad economy, the probability of simultaneous defaults is larger than usual. On the other hand, if the investor arranges his credit portfolio wisely, he/she might be able to lower the risk of simultaneous defaults, similar to the portfolio optimization on stocks, as discussed in section 2.1.3. However, this is only feasible if he/she can quantify the credit portfolio risk properly, which motivates this study.

The probability of losses can be described by the loss distribution, or the *loss given default* distribution. The latter distribution is based on the assumption that a default already occurred, whereas the default probability is already included in the loss distribution. This corresponds to a delta peak at $L = 0$ which, i.e., to the case of non-default. A typical loss distribution is illustrated in Fig. 4.1.

4.2 Types of Credit Risk Models

We can distinguish three fundamentally different types of credit risk models (see, e.g., Ref. [131]). The models of the first type are referred to as *structural models.* These models have a long history, going back to the work of Black and Scholes [19] and Merton [132]. The *Merton models* assume a zero-coupon debt structure with a fixed time to maturity T. The equity of the company is modeled by a stochastic process, for example describing the price of the company's stock. Thus, it can be seen as a creditor's call option on the obligor's assets. The risk of default and the associated *recovery rate*, the residual payment in case of a loss, are modeled by the company's equity at maturity.

The second type of model are *reduced-form models.* They are frameworks of many common macroscopic observables or risk factors. Despite of the large number of input parameters, their functional dependency is usually simple. Default risk and recovery rate are modeled independently. Some well known reduced-form model approaches can be found in Refs. [133–137]. These models have to be calibrated with current market data in order to give reliable results. They are commonly used in practice throughout the whole financial industry. From a physicist's point of view, these models are not satisfactory because due the large number of observables, they do not allow to gain fundamental insight into the mechanisms of a credit portfolio.

First passage models represent the third type of credit risk models. They are a mixed form of the previous two models and were first introduced by Black and Cox [138, 139]. Similar to Merton's model, the market value of a company is modeled through a stochastic process. However, a default occurs whenever the asset goes below a certain threshold for the first time. The recovery rates are typically modeled independently, for example, by a reduced-form model [140, 141] or are even assumed to be constant (see, e.g., Ref. [131]). Recent approaches aim at improving these models by including the chance of recovery, even if a company's market value is below the threshold [139] and estimating correlations between default probabilities of industry sectors [142].

The latter two models are often implemented in a computer software, for example, *CreditMetrics* by JP Morgan [143], *CreditPortfolioView* by McKinsey & Company [144] or *CreditRisk+* by Credit Suisse [145].

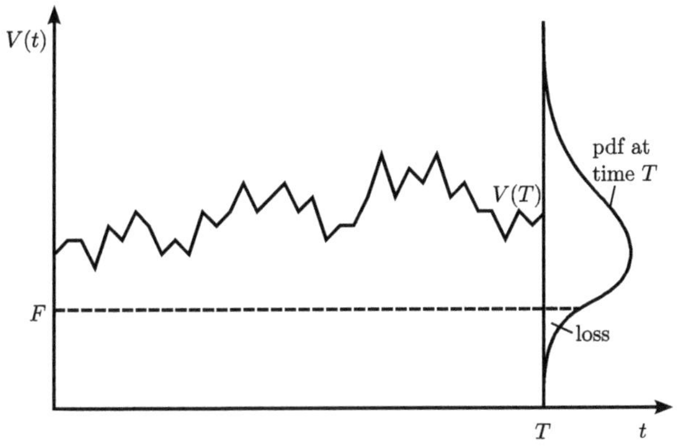

Figure 4.2: Credit process of the structural model.

As there can be a strong connection between default risks and recovery rates, the chances of large losses are often underestimated in the reduced-form and first passage models. Structural models do not require this separation.

Structural models provide a "microscopical" tool to study credit risk as the defaults and recoveries are traced back to stochastic processes modeling the state of individual obligors. Despite the high level of abstraction, there is a direct application of these models, namely speculative margin loans: A broker usually grants investors a credit based on his portfolio, a speculative margin loan. For example, if the value of the investor's portfolio value is USD 100 000, the broker might grant him a credit of USD 30 000 based on this portfolio. The investor can invest the money on more stocks and try to make more profit than the interest rate on this credit. As the broker maintains his portfolio, he can always take the investor's stocks as a security, if the investor is not willing to pay back his debt. However, if the value of the portfolio falls below USD 30 000 (plus interest), the broker is not able to get back the full amount he lent. A default occurs. Hence, structural models do not only give fundamental insights but also can have high practical relevance. In the next sections, we will develop a structural model with consideration for correlation.

4.3 A Structural Credit Risk Model

We develop a structural credit risk model step-by-step. Our model is based on Merton's original model. The aim of our model is to analytically describe the impact of correlations on the losses of a credit portfolio. In our model, a default occurs if the price V_k of the k-th asset is below the face value F_k at maturity time T. The size of the loss then depends on how far $V_k(T)$ is below the face value F_k. The situation is sketched in Fig. 4.2. The upright curve at time T represents the probability distribution, i.e., the pdf, of the price $V_k(T)$ at time T. The area below F_k in this curve represents the exposure to default risk. The motivation of the following is to calculate this area and consequently the loss distribution for a portfolio of K assets. We assume that the prices follow a geometric Brownian motion. As already stated in the introduction, there is a good agreement with empirical data on large timescales. Because the contract periods of credits are usually several years, this assumption is well justified.

4.3.1 Definitions and Conventions

Before we turn to the development of the model, we introduce some common notations that are used in the following sections.

4.3.1.1 Modeling of Asset Prices

As a short example for the notation used in this chapter, we turn to Noh's model, as already introduced in section 3.4 and extend it for multiple assets. To simplify matters, we assume a discrete time with time step $\Delta t = 1$. We can write a single increment of a Brownian motion with the variance σ_k for asset k as

$$\Delta V_k(t) = \sigma_k(\sqrt{1-c}\varepsilon_k(t) + \sqrt{c}\eta(t)) \ , \tag{4.1}$$

where ε_k and η are random variables normalized to unit variance. While ε_k is drawn individually for each k, all assets share the same $\eta(t)$ each time step.

The time series features the constant variance σ_k^2. A matrix notation is very convenient when modeling multiple price movements, i.e., a portfolio of K assets. As an example, we use K securities, all equally correlated with the coefficient c. In matrix notation ε in Eq. 4.1 corresponds to the vector

$\vec{B}(t)$, given by

$$\vec{B}(t) = \begin{pmatrix} \varepsilon_1(t) \\ \vdots \\ \varepsilon_K(t) \\ \eta(t) \end{pmatrix} . \tag{4.2}$$

$\vec{B}(t)$ consists of K components, corresponding to each asset plus one component, $\eta(t)$, that induces the correlation. Now we introduce the matrix $\mathbf{A}$ which holds the correlation coefficients and variances. For our example, this matrix $\mathbf{A}$ is given by a $K \times K+1$ matrix,

$$\mathbf{A} = \begin{pmatrix} \sigma_1\sqrt{1-c} & 0 & \cdots & 0 & \sigma_1\sqrt{c} \\ 0 & \ddots & & \vdots & \vdots \\ \vdots & & \ddots & 0 & \vdots \\ 0 & \cdots & 0 & \sigma_K\sqrt{1-c} & \sigma_K\sqrt{c} \end{pmatrix} . \tag{4.3}$$

$\mathbf{A}\mathbf{A}^\dagger$ leads to the covariance matrix $\mathbf{\Sigma}$,

$$\mathbf{A}\mathbf{A}^\dagger = \begin{pmatrix} \sigma_1^2 & c\sigma_1\sigma_2 & \cdots & c\sigma_1\sigma_K \\ c\sigma_2\sigma_1 & \ddots & & \vdots \\ \vdots & & \ddots & c\sigma_{K-1}\sigma_K \\ c\sigma_K\sigma_1 & \cdots & c\sigma_K\sigma_{K-1} & \sigma_K^2 \end{pmatrix} = \mathbf{\Sigma} . \tag{4.4}$$

If we multiply $\mathbf{A}$ and $\vec{B}(t)$, we obtain the vector of price increments of all assets K. For a single asset k, the component-wise notation reads

$$(\Delta\vec{V}(t))_k = \left(\mathbf{A}\vec{B}(t)\right)_k . \tag{4.5}$$

In this simple example of Noh's model, we only have $R = 1$ additional risk elements that corresponds to the market correlation. In more realistic scenarios, for example, when modeling several branches, we can have a larger number of risk elements R. The matrix $\mathbf{A}$ always consists of a square $K \times K$ matrix, stacked with R vectors that model the additional risk elements. With $K' = K + R$ total number of columns, $\mathbf{A}$ represents a $K \times K'$ matrix. Furthermore, for a simpler notation we introduce the $T \times N$ matrix $\mathbf{B}$ which consists of the vectors $\vec{B}(t)$ from $t = 1$ to $t = T$ as columns. We will use this notation consisting of $\mathbf{A}$, $\mathbf{B}$ and $\vec{B}(t)$ throughout this chapter. Moreover, we do not restrict ourselves to the aforementioned example of Noh's model. The statistical modeling of asset prices can also be more complex.

4.3.1.2 Multivariate Gaussian Distributions

The distribution of a vector $\vec{v} \in \mathbb{R}^N$ with independently Gaussian distributed entries v_i, can be written with a scalar product,

$$p(\vec{v}) = \left(\frac{1}{\sqrt{2\pi}}\right)^N \prod_{i=1}^{N} \left(\frac{1}{\sigma_i} \exp\left(-\frac{(v_i - \mu_i)^2}{2\sigma_i^2}\right)\right) \tag{4.6}$$

$$= \underbrace{\left(\frac{1}{\sqrt{2\pi}}\right)^N \prod_{i=1}^{N} \left(\frac{1}{\sigma_i}\right)}_{C_{\text{norm}}} \exp\left(-\frac{1}{2}\sum_{i=1}^{N}\Big(\overbrace{\frac{v_i - \mu_i}{\sigma_i}}^{\hat{v}_i}\Big)^2\right) \tag{4.7}$$

$$= C_{\text{norm}} \exp\left(-\frac{1}{2}\vec{\hat{v}}^\dagger \vec{\hat{v}}\right) , \tag{4.8}$$

where $\vec{\mu}$ refers to the vector of expectation values and $\vec{\sigma}$ is the vector of standard deviations. C_{norm} is the normalization factor and $\hat{v}_i$ refers to the normalized components of the vector $\vec{v}$. If the components are standard normal distributed, the distribution can be written as

$$p(\vec{v}) = \left(\frac{1}{\sqrt{2\pi}}\right)^N \exp\left(-\frac{1}{2}\vec{v}^\dagger \vec{v}\right) . \tag{4.9}$$

Analogously, the distribution of a matrix $\mathbf{W} \in \mathbb{R}^{N\times M}$ with Gaussian distributed components W_{ij} can be written using a trace,

$$p(\mathbf{W}) = \left(\frac{1}{\sqrt{2\pi}}\right)^{NM} \prod_{i=1}^{N}\prod_{j=1}^{M} \left(\frac{1}{\sigma_{ij}} \exp\left(-\frac{(W_{ij} - \mu_{ij})^2}{2\sigma_{ij}^2}\right)\right) \tag{4.10}$$

$$= \underbrace{\left(\frac{1}{\sqrt{2\pi}}\right)^{NM} \prod_{i=1}^{N}\prod_{j=1}^{M} \left(\frac{1}{\sigma_{ij}}\right)}_{C_{\text{norm}}} \exp\left(-\frac{1}{2}\sum_{i=1}^{N}\sum_{j=1}^{M}\Big(\overbrace{\frac{W_{ij} - \mu_{ij}}{\sigma_{ij}}}^{\widehat{W}_{ij}}\Big)^2\right) \tag{4.11}$$

$$= C_{\text{norm}} \exp\left(-\frac{1}{2}\sum_{k=1}^{N}(\widehat{\mathbf{W}}\widehat{\mathbf{W}}^\dagger)_{kk}\right) \tag{4.12}$$

$$= C_{\text{norm}} \exp\left(-\frac{1}{2}\text{tr}(\widehat{\mathbf{W}}\widehat{\mathbf{W}}^\dagger)\right) . \tag{4.13}$$

And for the case of standard normal distributed entries,

$$p(\mathbf{W}) = \left(\frac{1}{\sqrt{2\pi}}\right)^{NM} \exp\left(-\frac{1}{2}\mathrm{tr}(\mathbf{W}\mathbf{W}^\dagger)\right) . \tag{4.14}$$

We will use this very convenient notation in the following sections.

4.3.1.3 Multidimensional Integration

We will often integrate over all entries of a matrix. Thus we introduce the following abbreviation. When integrating over all elements of the $T \times N$ matrix $\mathbf{B}$, we write

$$d[\mathbf{B}] \equiv \prod_{t=1}^{T}\prod_{n=1}^{N} dB_{nt} . \tag{4.15}$$

Analogously, we write the integration over the elements of a N dimensional vector by

$$d[\vec{B}] \equiv \prod_{n=1}^{N} dB_n . \tag{4.16}$$

4.3.2 Price Distribution

As we now have the prerequisites, we can develop a description for the evolution of the asset prices' probability density. First, we develop our model for a Brownian motion, which we will later extend to a geometric Brownian motion. In both cases, the evolution of prices depends on the covariance matrix $\mathbf{\Sigma}$. In a next step, we assume that this matrix has random entries. Thus, we will develop the average distribution of prices under the assumption of random correlations with average correlation level zero.

4.3.2.1 Brownian Motion

To simplify matters, let us first consider the case of a Brownian motion without drift. In order to obtain the distribution of the asset prices after time T, we integrate over all probability densities $p(\mathbf{B})$ of the random variables in $\mathbf{B}$ and filter for those combinations of the random variables, which satisfy the Brownian motion,

$$p(\vec{V}(T), \mathbf{A}, \mathbf{B}) = \int p(\mathbf{B})\delta\left(\vec{V}(T) - \sum_{t=1}^{T}(\mathbf{A}\vec{B}(t))\right) d[\mathbf{B}] . \tag{4.17}$$

In this notation, the distribution depends on the model setup variables **A** and **B**. The aim of the following calculation is to change this dependency to the covariance matrix, **Σ**. The covariance matrix can be measured directly, while **A** and **B** are constructions of the model. We first turn to the delta distribution in Eq. (4.17). Using a Fourier transform, we can write the delta distribution as

$$\delta\left(V(\vec{T}) - \sum_{t=1}^{T} \mathbf{A}\vec{B}(t)\right) \tag{4.18}$$

$$= \prod_{k=1}^{K} \delta\left(V_k(T) - \sum_{t=1}^{T} \left(\mathbf{A}\vec{B}(t)\right)_k\right) \tag{4.19}$$

$$= \left(\frac{1}{2\pi}\right)^K \prod_{k=1}^{K} \int_{-\infty}^{+\infty} \exp\left(-i\omega_k V_k(T) + i\omega_k \sum_{t=1}^{T} \left(\mathbf{A}\vec{B}(t)\right)_k\right) d\omega_k \tag{4.20}$$

$$= \left(\frac{1}{2\pi}\right)^K \int \exp\left(-i\vec{\omega}^\dagger \vec{V}(T)\right) \exp\left(i\vec{\omega}^\dagger \sum_{t=1}^{T} \mathbf{A}\vec{B}(t)\right) d[\vec{\omega}]\ . \tag{4.21}$$

In the last step, we changed the integration over all ω_k to the whole vector $\vec{\omega}$. Now we insert (4.21) into the initial distribution of asset prices (4.17), change the order of integration and obtain

$$p(\vec{V}(T), \mathbf{A}, \mathbf{B}) = \left(\frac{1}{2\pi}\right)^K \int \exp\left(-i\vec{\omega}^\dagger \vec{V}(T)\right) \times \underbrace{\prod_{t=1}^{T} \int p(\vec{B}(t)) \exp\left(i\vec{\omega}^\dagger \mathbf{A}\vec{B}(t)\right) d[\vec{B}(t)]}_{R(\omega)} d[\vec{\omega}]\ . \tag{4.22}$$

Here, we changed the initial integration over all entries of the matrix **B** to a product of intervals of all entries of the vector $\vec{B}(t)$. We identify $R(\omega)$ as the characteristic function of this distribution. Now we insert $p(\vec{B}(t))$ as a multivariate standard normal distribution (the individual variances are modeled by **A**),

$$p(\vec{B}(t)) = \left(\frac{1}{\sqrt{2\pi}}\right)^N \exp\left(-\frac{1}{2}\vec{B}^\dagger(t)\vec{B}(t)\right) \tag{4.23}$$

and perform a further Fourier transformation,

$$R(\omega) = \left(\frac{1}{\sqrt{2\pi}}\right)^{NT} \prod_{t=1}^{T} \int \exp\left(-\frac{1}{2}\vec{B}(t)^{\dagger}\vec{B}(t)\right) \times \exp\left(i\vec{\omega}^{\dagger}\mathbf{A}\vec{B}(t)\right) d[\vec{B}(t)] \tag{4.24}$$

$$= \left(\frac{1}{\sqrt{2\pi}}\right)^{NT} \prod_{t=1}^{T} \sqrt{2\pi}^{N} \exp\left(-\frac{1}{2}\vec{\omega}^{\dagger}\mathbf{A}\mathbf{A}^{\dagger}\vec{\omega}\right) \tag{4.25}$$

$$= \exp\left(-\frac{T}{2}\vec{\omega}^{\dagger}\mathbf{\Sigma}\vec{\omega}\right) . \tag{4.26}$$

If we insert this result in Eq. (4.22), we have

$$p(\vec{V}(T), \mathbf{\Sigma}) = \left(\frac{1}{2\pi}\right)^{K} \int \exp\left(-i\vec{\omega}^{\dagger}\vec{V}(T)\right) \exp\left(-\frac{T}{2}\vec{\omega}^{\dagger}\mathbf{\Sigma}\vec{\omega}\right) d[\vec{\omega}] , \tag{4.27}$$

where the integrand is the characteristic function of a multivariate Gaussian distribution. Thus, with a further Fourier transformation, we obtain the well known result of a multivariate Gaussian distribution,

$$p(\vec{V}(T), \mathbf{\Sigma}) = \left(\frac{1}{\sqrt{2\pi T}}\right)^{K} \frac{1}{\sqrt{\det(\mathbf{\Sigma})}} \exp\left(-\frac{1}{2T}\vec{V}(T)^{\dagger}\mathbf{\Sigma}^{-1}\vec{V}(T)\right) . \tag{4.28}$$

We can easily extend this result by introducing a constant drift, described by the vector $\vec{\mu}$. Eq. (4.17) then becomes

$$p(\vec{V}(T), \mathbf{A}, \mathbf{B}) = \prod_{t=1}^{T} \int p(\vec{B}(t)) \times \delta\left(\vec{V}(T) - \vec{\mu}T - \sum_{t=1}^{T}(\mathbf{A}\vec{B}(t))\right) d[\vec{B}(t)] . \tag{4.29}$$

This leads to the result

$$p(\vec{V}(T), \mathbf{\Sigma}) = \left(\frac{1}{\sqrt{2\pi T}}\right)^{K} \frac{1}{\sqrt{\det(\mathbf{\Sigma})}} \times \exp\left(-\frac{1}{2T}(\vec{V}(T) - \vec{\mu}T)^{\dagger}\mathbf{\Sigma}^{-1}(\vec{V}(T) - \vec{\mu}T)\right) . \tag{4.30}$$

4.3.2.2 Geometric Brownian Motion

In the previous part, we developed the distribution of prices $\vec{V}$ under the assumption of a Brownian motion. However, a geometric Brownian motion is the suitable process for the modeling of asset prices. In the following, we will show that this formalism can easily be extended to a geometric Brownian motion.

We can map the case of a geometric Brownian motion to an ordinary Brownian motion (see, e.g., Ref. [146]). This is accomplished by substituting in Eq. (4.17),

$$V_k(T) \to \widehat{V}_k(T) = \ln\left(\frac{V_k(T)}{V_{k,0}}\right) - \left(\mu_k - \frac{\sigma_k^2}{2}\right)T \tag{4.31}$$

leading to

$$\begin{aligned} p(\vec{V}(T), \mathbf{A}, \mathbf{B}) = & \prod_{t=1}^{T}\prod_{k=1}^{K} \frac{1}{V_k(T)} \int p_k(\vec{B}(t)) \\ & \times \delta\left(\widehat{V}_k(T) - \sum_{t=1}^{T}((\mathbf{A}\vec{B}(t))_k\right) d[\vec{B}(t)] \end{aligned} \tag{4.32}$$

and thus eventually leading to

$$\begin{aligned} p(\vec{V}(T), \mathbf{\Sigma}) = & \frac{1}{\prod_{k=1}^{K} V_k(T)} \left(\frac{1}{\sqrt{2\pi T}}\right)^K \frac{1}{\sqrt{\det(\mathbf{\Sigma})}} \\ & \times \exp\left(-\frac{1}{2T}\widehat{\vec{V}}(T)^{\dagger}\mathbf{\Sigma}^{-1}\widehat{\vec{V}}(T)\right) . \end{aligned} \tag{4.33}$$

This is the probability density of the prices at time T implying a geometric Brownian motion with the covariance matrix $\mathbf{\Sigma}$. The factors $1/V_k$ are a consequence of the substitution in Eq. 4.31. They ensure proper normalization.

4.3.3 Average Price Distribution

With Eq. (4.33) we have found an expression for the probability distribution of asset prices in the case of a correlated Brownian motion. However, we are not interested in the impact of a specific correlation matrix. Instead we want to estimate the general impact of correlations. To this end, we want to average over all possible correlation matrices and disclose the general statistical behavior of the system. This will enable us to make a profound statistical statement.

We use a random matrix approach to calculate the average price distribution for random correlations where the average correlation level is zero. By averaging over all possible combinations of random variables, we obtain the average price distribution $\langle p(\vec{V}(T))\rangle$ under these assumptions.

Thus, we replace the covariance matrix $\mathbf{\Sigma} = \mathbf{A}\mathbf{A}^\dagger$ of the previous section with a random correlation matrix and the matrix of standard deviations $\mathbf{S} = \mathrm{diag}(\sigma_1, \ldots, \sigma_K)$

$$\mathbf{\Sigma}_W = \mathbf{S}\mathbf{W}\mathbf{W}^\dagger\mathbf{S} \ , \tag{4.34}$$

where $\mathbf{W} \in \mathbb{R}^{K\times N}$ is a random matrix. Hence, the element $(\Sigma_W)_{ij}$ consists of a random number multiplied by $\sigma_i\sigma_j$. The entries of $\mathbf{W}$ are Gaussian distributed with

$$p(\mathbf{W}) = \left(\sqrt{\frac{N}{2\pi}}\right)^{KN} \exp\left(-\frac{N}{2}\mathrm{tr}\left(\mathbf{W}\mathbf{W}^\dagger\right)\right) \ . \tag{4.35}$$

Here, we add the factor N in the nominator because we need the variance in the average price distribution to be independent from N. We will demonstrate that this supplement leads to N-independence later in this section. For $N \to \infty$, we obtain the probability density function of a unit matrix. This represents the uncorrelated case. Since we only consider correlation matrices with full rank, we obtain the strongest correlations if we choose $N = K$. The case $N < K$ is disregarded as the resulting matrix is not invertible which is usually required for applications in risk management. When inserting this ansatz in Eq. (4.27), we obtain

$$\langle p(\vec{V}(T))\rangle = \int p(\mathbf{W}) p(\vec{V}(T), \mathbf{S}\mathbf{W}\mathbf{W}^\dagger\mathbf{S}) d[\mathbf{W}] \tag{4.36}$$

$$= \left(\frac{1}{\sqrt{2\pi}}\right)^{2K+KN} \sqrt{N}^{KN} \int \exp\left(-\frac{N}{2}\mathrm{tr}\left(\mathbf{W}\mathbf{W}^\dagger\right)\right)$$
$$\times \int \exp\left(-i\vec{\omega}^\dagger \vec{V}(T)\right) \exp\left(-\frac{T}{2}\vec{\omega}^\dagger \mathbf{S}\mathbf{W}\mathbf{W}^\dagger\mathbf{S}\vec{\omega}\right) d[\vec{\omega}] d[\mathbf{W}] \tag{4.37}$$

$$
\begin{aligned}
&= \left(\frac{1}{\sqrt{2\pi}}\right)^{2K+KN} \sqrt{N}^{KN} \int \exp\left(-i\vec{\omega}^\dagger \vec{V}(T)\right) \\
&\quad \times \int \exp\left(-\frac{N}{2}\mathrm{tr}\left(\mathbf{W}\mathbf{W}^\dagger\right)\right) \\
&\quad \times \exp\Big(-\frac{T}{2}\overbrace{\vec{\omega}^\dagger \mathbf{S}\mathbf{W}\mathbf{W}^\dagger \mathbf{S}\vec{\omega}}^{=\mathrm{tr}(\vec{\omega}^\dagger \mathbf{S}\mathbf{W}\mathbf{W}^\dagger \mathbf{S}\vec{\omega})}\Big) d[\mathbf{W}]d[\vec{\omega}]\,. \qquad (4.38)
\end{aligned}
$$

In the last steps, we took advantage of the fact that the term $\vec{\omega}^\dagger \mathbf{S}\mathbf{W}\mathbf{W}^\dagger \mathbf{S}\vec{\omega}$ results in a scalar, which can be written as its trace. As the trace is invariant in cyclic permutation, we can express this term as $\mathrm{tr}(\mathbf{W}\mathbf{W}^\dagger \mathbf{S}\vec{\omega}\vec{\omega}^\dagger \mathbf{S})$. Hence, we write with $\mathbf{I}$ denoting the identity matrix,

$$
\begin{aligned}
\langle p(\vec{V}(T))\rangle &= \left(\frac{1}{\sqrt{2\pi}}\right)^{2K+KN} \sqrt{N}^{KN} \int \exp\left(-i\vec{\omega}^\dagger \vec{V}(T)\right) \\
&\quad \times \int \exp\left(-\frac{1}{2}\mathrm{tr}(\mathbf{W}\mathbf{W}^\dagger(\mathbf{I}N + T\mathbf{S}\vec{\omega}\vec{\omega}^\dagger \mathbf{S}))\right) d[\mathbf{W}]d[\vec{\omega}] \qquad (4.39) \\
&= \left(\frac{1}{\sqrt{2\pi}}\right)^{2K+KN} \sqrt{N}^{KN} \int \exp\left(-i\vec{\omega}^\dagger \vec{V}(T)\right) \\
&\quad \times \int \exp\left(-\frac{1}{2}\sum_{n=1}^{N}(\vec{W}_n^\dagger(\mathbf{I}N + T\mathbf{S}\vec{\omega}\vec{\omega}^\dagger \mathbf{S})\vec{W}_n)\right) d[\vec{W}_n]d[\vec{\omega}] \qquad (4.40) \\
&= \left(\frac{1}{\sqrt{2\pi}}\right)^{2K+KN} \sqrt{N}^{KN} \int \exp\left(-i\vec{\omega}^\dagger \vec{V}(T)\right) \\
&\quad \times \left(\int \exp\Big(-\frac{1}{2}\overbrace{\vec{W}^\dagger}^{\equiv x_i}\overbrace{(\mathbf{I}N + T\mathbf{S}\vec{\omega}\vec{\omega}^\dagger \mathbf{S})}^{\equiv A_{ij}}\overbrace{\vec{W}}^{\equiv x_j}\Big) d[\vec{W}]\right)^N d[\vec{\omega}]\,. \qquad (4.41)
\end{aligned}
$$

The last step can be accomplished, as the components of $\mathbf{W}$ are independent identically distributed, hence we can denote the n-th column vector of $\mathbf{W}$, $\vec{W}_n$ by $\vec{W}$. Thus, we can simplify the integration over the matrix $\mathbf{W}$ to the integration over the vector $\vec{W} \in \mathbb{R}^K$ to the power of N.

The integral over $d[\vec{W}]$ is simply a Gaussian integral, as indicated by x_i, x_j and A_{ij}. As $\vec{W}_n$ consists of K components, this gives an additional factor

$\sqrt{2\pi}^{KN}$ and thus leads to

$$\langle p(\vec{V}(T))\rangle = \left(\frac{\sqrt{N}^N}{2\pi}\right)^K \int \exp\left(-i\vec{\omega}^\dagger \vec{V}(T)\right) \frac{1}{\sqrt{\det(\mathbf{I}N + T\mathbf{S}\vec{\omega}\vec{\omega}^\dagger\mathbf{S})}^N} d[\vec{\omega}] \ . \tag{4.42}$$

The determinant can also be expressed as

$$\det(\mathbf{I}N + T\mathbf{S}\vec{\omega}\vec{\omega}^\dagger\mathbf{S}) = N^K \left(1 + \frac{T}{N}\vec{\omega}^\dagger\mathbf{S}\mathbf{S}\vec{\omega}\right) \tag{4.43}$$

(see appendix B.1). Hence, we arrive at

$$\langle p(\vec{V}(T))\rangle = \left(\frac{1}{2\pi}\right)^K \int \exp\left(-i\vec{\omega}^\dagger \vec{V}(T)\right) \frac{1}{(1 + (T/N)\vec{\omega}^\dagger\mathbf{S}\mathbf{S}\vec{\omega})^{N/2}} d[\vec{\omega}] \ . \tag{4.44}$$

This integral can be solved by identifying the Gamma function [147], as

$$\frac{\Gamma(x)}{a^x} = \int_0^\infty z^{x-1} \exp\left(-az\right) dz \quad , \quad x > 0, \ a > 0 \ . \tag{4.45}$$

We identify a^{-x} with $((1 + (T/N)\vec{\omega}^\dagger\mathbf{S}\mathbf{S}\vec{\omega}))^{-N/2}$ and obtain

$$\begin{aligned}
\langle p(\vec{V}(T))\rangle &= \left(\frac{1}{2\pi}\right)^K \frac{1}{\Gamma(N/2)} \int \exp\left(-i\vec{\omega}^\dagger \vec{V}(T)\right) \\
&\quad \times \int_0^\infty z^{\left(\frac{N}{2}-1\right)} \exp\left(-(1 + \frac{T}{N}\vec{\omega}^\dagger\mathbf{S}\mathbf{S}\vec{\omega})z\right) dz \, d[\vec{\omega}] && (4.46) \\
&= \left(\frac{1}{2\pi}\right)^K \frac{1}{\Gamma(N/2)} \int_0^\infty z^{\left(\frac{N}{2}-1\right)} \exp\left(-z\right) \\
&\quad \times \int \exp\left(-i\vec{\omega}^\dagger \vec{V}(T)\right) \exp\left(-\frac{T}{N}\sum_{k=1}^K \sigma_k^2\omega_k^2 z\right) d[\vec{\omega}] dz && (4.47) \\
&= \left(\frac{1}{2\pi}\right)^K \frac{1}{\Gamma(N/2)} \int_0^\infty z^{\left(\frac{N}{2}-1\right)} \exp\left(-z\right)
\end{aligned}$$

$$\times \prod_{k=1}^{K} \left[\int \exp\left(-i\omega_k V_k(T)\right) \exp\left(-\frac{T}{N}\sigma_k^2\omega_k^2 z\right) d\omega_k \right] dz \quad (4.48)$$

$$= \left(\frac{1}{2\pi}\right)^K \frac{1}{\Gamma(N/2)} \int_0^\infty z^{\left(\frac{N}{2}-1\right)} \exp\left(-z\right)$$

$$\times \prod_{k=1}^{K} \left[\frac{\sqrt{\pi N}}{\sqrt{zT}\sigma_k} \exp\left(-\frac{NV_k^2}{4Tz\sigma_k^2}\right) \right] dz \quad (4.49)$$

$$= \left(\frac{1}{2\pi}\right)^K \frac{1}{\Gamma(N/2)} \left(\prod_{k=1}^{K} \frac{1}{\sigma_k}\right)$$

$$\times \int_0^\infty z^{\left(\frac{N}{2}-1\right)} \exp\left(-z\right) \left(\frac{\pi N}{zT}\right)^{\frac{K}{2}} \exp\left(-\frac{N}{4Tz}\sum_{k=1}^{K} \frac{V_k^2}{\sigma_k^2}\right) . \quad (4.50)$$

This integral is a representation of the Bessel function of the second kind $\mathcal{K}$ of the order $(K-N)/2$ [148]. Thus, we obtain the result

$$\langle p(\vec{V}(T)) \rangle = \left(\sqrt{\frac{N}{2\pi T}}\right)^K \frac{1}{\Gamma(N/2)} \left(\prod_{k=1}^{K} \frac{1}{\sigma_k}\right)$$

$$\times 2^{1-\frac{N}{2}} \left(\sqrt{\frac{N}{T}\sum_{k=1}^{K} \frac{V_k^2}{\sigma_k^2}}\right)^{\frac{N-K}{2}} \mathcal{K}_{\frac{K-N}{2}} \left(\sqrt{\frac{N}{T}\sum_{k=1}^{K} \frac{V_k^2}{\sigma_k^2}}\right) . \quad (4.51)$$

This is the average distribution of $p(\vec{V}(T))$ if implying a randomly distributed correlation matrix and an average correlation level of zero. We stated earlier in this chapter, that we include N in the distribution of the random matrices $\mathbf{W}$ in order to render the variance of the average price distribution N-independent. The variances only depend on T and σ_k. The parameter N is only used to control the correlations. We can demonstrate this by calculating the variance of the i-th price V_i,

$$\mathrm{var}(V_i(T)) = \int d[V] \; V_i(T)^2 \langle p(\vec{V}(T)) \rangle \; . \quad (4.52)$$

We can solve this integral by using hyperspherical coordinates.

We chose the length

$$\rho \equiv \sqrt{\sum_{k=1}^{K} \frac{V_k^2}{\sigma_k^2}} \tag{4.53}$$

and parametrize the entry with fixed index i according to

$$\frac{V_i}{\sigma_i} \equiv \rho \cos(\vartheta) \ . \tag{4.54}$$

Now we can write the integral in Eq. (4.52) as

$$\begin{aligned} \mathrm{var}(V_i(T)) = {} & \left(\sqrt{\frac{N}{2\pi T}}\right)^K \frac{1}{\Gamma(N/2)} \left(\prod_{k=1}^{K} \frac{1}{\sigma_k}\right) 2^{1-\frac{N}{2}} \int_0^\infty d\rho \rho^{K-1} \\ & \times \int_0^\pi d\vartheta \sin(\vartheta)^{K-2} \sigma_i^2 \rho^2 \cos(\vartheta)^2 \left(\sqrt{\frac{N}{T}}\rho\right)^{\frac{N-K}{2}} \mathcal{K}_{\frac{K-N}{2}}\left(\sqrt{\frac{N}{T}}\rho\right) \\ & \times \int d\Omega_{k-1} \ , \end{aligned} \tag{4.55}$$

where the proper sphere in $K-1$ dimensions has the volume

$$\int d\Omega_{K-1} = \frac{2\pi^{(K-1)/2}}{\Gamma((K-1)/2)} \prod_{k=1}^{K} \sigma_k \tag{4.56}$$

(as worked out in appendix B.2). Hence, we obtain

$$\mathrm{var}(V_i(T)) = 2\sigma_i^2 \frac{T}{N} \frac{\Gamma(N/2+1)}{\Gamma(N/2)} \tag{4.57}$$

$$= \sigma_i^2 T \ . \tag{4.58}$$

Thus, the variance of V_i only depends on the standard deviation σ_i and the time T. Analogously, in case of a geometric Brownian motion by a simple substitution as in Eq. (4.31) we obtain,

$$\begin{aligned} \langle p(\vec{V}(T)) \rangle = {} & \left(\sqrt{\frac{N}{2\pi T}}\right)^K \frac{1}{\Gamma(N/2)} \left(\prod_{k=1}^{K} \frac{1}{\sigma_k V_k}\right) \\ & \times 2^{1-\frac{N}{2}} \left(\sqrt{\frac{N}{T} \sum_{k=1}^{K} \frac{\widehat{V}_k^2}{\sigma_k^2}}\right)^{\frac{N-K}{2}} \mathcal{K}_{\frac{K-N}{2}}\left(\sqrt{\frac{N}{T} \sum_{k=1}^{K} \frac{\widehat{V}_k^2}{\sigma_k^2}}\right) \end{aligned} \tag{4.59}$$

with

$$\widehat{V}_k(T) = \ln\left(\frac{V_k(T)}{V_{k,0}}\right) - \left(\mu_k - \frac{\sigma_k^2}{2}\right) T \,. \tag{4.60}$$

Here, parameter σ_k refers to the standard deviation of the underlying Brownian motion, i.e., the volatility of asset returns. The resulting prices thus have the standard deviation,

$$\hat{\sigma}_k = \sqrt{\exp\left(2\mu + \sigma_k^2 T\right)\left(\exp\left(\sigma_k^2 T\right) - 1\right) V_{k,0}^2} \,. \tag{4.61}$$

Fig. 4.3 illustrates the probability density implying a Brownian motion, as given in Eq. (4.51) for the two-dimensional $K = 2$ case. For $N = K$, we obtain a very narrow, but heavy-tailed distribution. For larger values of N, the distribution slowly approaches to an uncorrelated bivariate Gaussian distribution. The standard deviation is identical in all three figures. The width in the center increases and the tails become weaker. Fig. 4.4 shows the distribution of prices based on a geometric Brownian motion, as given in Eq. (4.59). The findings are similar to the first case. While we obtain a narrow but heavy-tailed distribution for $N = K$, the distribution becomes more similar to an uncorrelated bivariate log-normal distribution with increasing values of N.

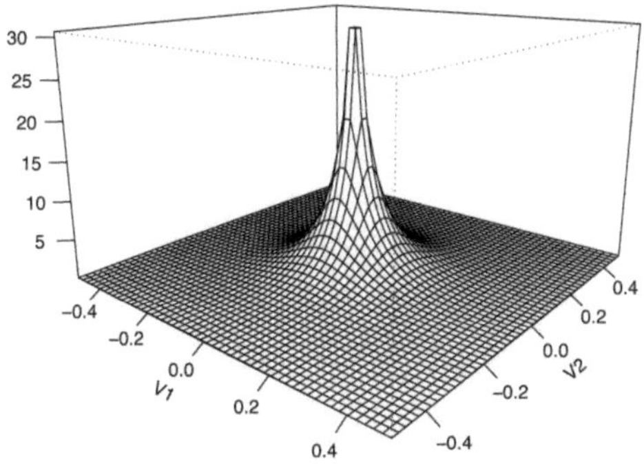

(a) N=2

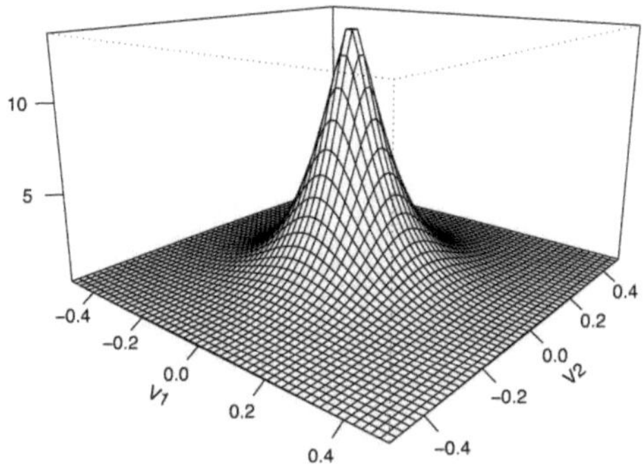

(b) N=4

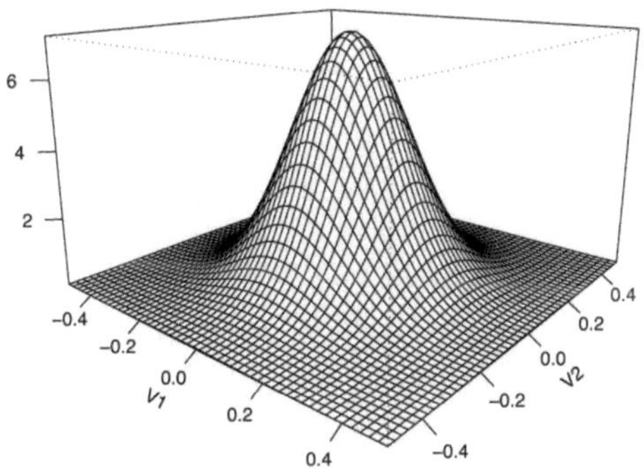

(c) N=100

Figure 4.3: Illustration of the average price distribution $\langle p(\vec{V}(T))\rangle$ assuming a Brownian motion for $T = 1$, $K = 2$ and different values for N. All distributions have the identical standard deviation $\sigma = 0.15$. For $N = 2$, we obtain a heavy-tailed distribution while with a singularity at the origin; the Gaussian limit is reached for $N = 100$.

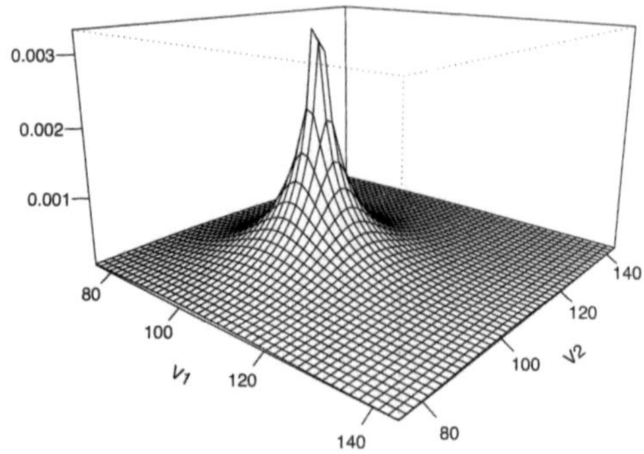

(a) N=2

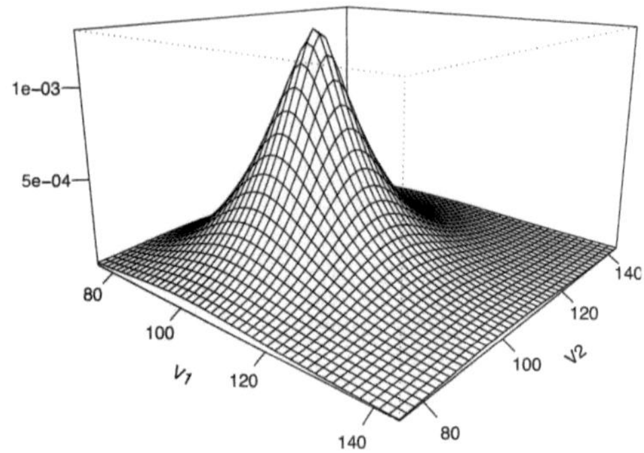

(b) N=4

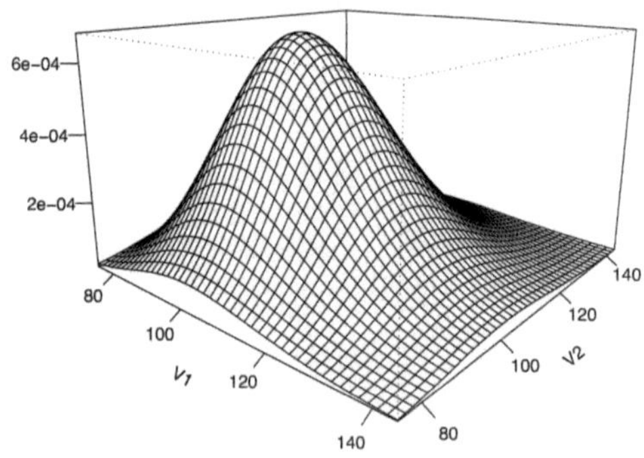

(c) N=100

Figure 4.4: Illustration of the average price distribution $\langle p(\vec{V}(T))\rangle$ assuming a geometric Brownian motion for $T = 1$, $K = 2$, $V_{k,0} = 100$, $\mu = 0.05$ and different values for N. All distributions have the identical standard deviation $\hat{\sigma} \approx 16$ ($\sigma = 0.15$). For $N = 2$, we obtain a heavy-tailed distribution while the uncorrelated limit is reached for $N = 100$.

4.3.4 Loss Distribution

We now turn to the calculation of the loss distribution. In our model, a default occurs if the price at maturity $V_k(T)$, corresponding to the obligor's equity, is below the face value F_k. The size of the loss is given by the difference of F_k and $V_k(T)$. Even if a loss occurs, the creditor might not lose all money that he lent, because the obligor is still able to pay back the amount $V_k(T)$. To compare losses in a portfolio of credits, we have to normalize them by the corresponding face value. We define the loss L_k of the k-th asset as

$$L_k = \begin{cases} \frac{F_k - V_k(T)}{F_k} & V_k(T) < F_k \qquad \text{(default)} \\ 0 & \text{else} \qquad \text{(no default)} \end{cases} . \tag{4.62}$$

When calculating the loss distributions, we have to take into account that in Eq. (4.62), the prices have to be positive. Therefore we assume in all further considerations that the underlying process of the price distribution follows a geometric Brownian motion.

When calculating the overall loss of a portfolio, we have to weight each loss by its face value in relation to the sum of all portfolio face values,

$$L = \sum_{k=1}^{K} f_k L_k \quad , \quad f_k = \frac{F_k}{\sum_{l=1}^{K} F_l} . \tag{4.63}$$

For example if one credit has a loss of 50% by Eq. (4.62) and a face value of USD 10 000 while another credit has a loss of 90% and a face value of USD 100 000, one cannot simply average the losses. A proper normalization of the face values gives an overall loss of $L \approx 86\%$.

The approach for the loss distribution follows the same principles as the approach for the price distribution in section 4.3.2. We integrate over the distribution of prices and filter for those that lead to a given total loss L. By the above stated definitions, we can define a filter for the total loss at maturity time T. In the next step we express the filter using a Fourier transformation. Eventually, we separate those terms that correspond to a default and those that describe the price above the face value F_k,

$$p(L) = \int_0^\infty d[\vec{V}(T)]p(\vec{V}(T))\delta\left(L - \sum_{k=1}^K f_k L_k\right) \tag{4.64}$$

$$= \int_0^\infty d[\vec{V}](T)p(\vec{V}(T))\frac{1}{2\pi}\int_{-\infty}^{+\infty} d\omega \exp\left(-i\omega L + i\omega \sum_{k=1}^K f_k L_k\right) \tag{4.65}$$

$$= \frac{1}{2\pi}\int_{-\infty}^{+\infty} d\omega \exp\left(-i\omega L\right)\int_0^\infty d[\vec{V}(T)] \exp\left(i\omega \sum_{k=1}^K f_k L_k\right) p(\vec{V}(T)) \tag{4.66}$$

$$= \frac{1}{2\pi}\int_{-\infty}^{+\infty} d\omega \exp\left(-i\omega L\right)$$
$$\times \left(\left(\int_0^{F_1} dV_1(T) \exp\left(i\omega f_1\left(1 - \frac{V_1}{F_1}\right)\right)\right) \times \dots\right.$$
$$\times \int_0^{F_K} dV_K(T) \exp\left(i\omega f_K\left(1 - \frac{V_K}{F_K}\right)\right)$$
$$\left. + \int_{F_k}^\infty dV_1(T) \times \dots \times \int_{F_k}^\infty dV_K(T)\right) p(\vec{V}(T)) \tag{4.67}$$

$$= \frac{1}{2\pi}\int_{-\infty}^{+\infty} d\omega \exp\left(-i\omega L\right)$$
$$\times \prod_{k=1}^K \left[\int_0^{F_k} dV_k(T) \exp\left(i\omega f_k\left(1 - \frac{V_k}{F_k}\right)\right) + \int_{F_k}^\infty dV_k\right] p(\vec{V}(T))\ . \tag{4.68}$$

Here, the expression in the squared brackets acts as an operator, because $p(V)$ does not necessarily factorize. We will use this ansatz to calculate the average loss distribution in the next section. However, Eq. (4.68) can be used to calculate the loss distribution if the actual price distribution is known, i.e., the statistical dependence and the underlying process are estimated. To prepare for this it is handy to write Eq. (4.68) as a combinatorial sum,

$$
\begin{aligned}
p(L) = \frac{1}{2\pi} \int\limits_{-\infty}^{+\infty} d\omega \exp\left(-i\omega L\right) \\
\times \sum_{k=1}^{K} \sum_{j=1}^{\binom{K}{k}} \Bigg[\prod_{l \in \mathrm{Perm}(j,k,K)} \int\limits_{0}^{F_l} dV_l \exp\left(i\omega f_l \left(1 - \frac{V_l}{F_l}\right)\right) \\
\times \prod_{\substack{q \in \{1 \dots K\} \\ \setminus \mathrm{Perm}(j,k,K)}} \int\limits_{F_q}^{\infty} dV_q \Bigg] p(\vec{V}(T)) \ ,
\end{aligned}
\tag{4.69}
$$

where $\mathrm{Perm}(j, k, K)$ is the j-th permutation of k elements of the set $\{1 \dots K\}$. For example, if $K = 3$ and $k = 2$, we obtain, $\mathrm{Perm}(1, 2, 3) = \{1, 2\}$, $\mathrm{Perm}(2, 2, 3) = \{2, 3\}$ and $\mathrm{Perm}(3, 2, 3) = \{1, 3\}$. However, Eq. (4.69) might need to be estimated numerically, depending on the complexity of the price distribution $p(\vec{V}(T))$. In section 4.3.6, we will simplify this combinatorial sum for a homogeneous portfolio and the average price distribution $\langle p(\vec{V}(T))\rangle$.

4.3.5 Average Loss Distribution

Now we have developed all necessary tools to model the average distribution of losses under the assumption of random correlations and an average correlation level of zero. We start by inserting the average price distribution (4.49) into the loss distribution (4.68),

$$
\begin{aligned}
\langle p(L)\rangle = \frac{1}{2\pi\Gamma(N/2)} \int\limits_{0}^{\infty} dz \; z^{\left(\frac{N}{2}-1\right)} \exp\left(-z\right) \int\limits_{-\infty}^{+\infty} d\omega \exp\left(-i\omega L\right) \\
\times R(\omega, z) \ ,
\end{aligned}
\tag{4.70}
$$

with

$$R(\omega, z) = \prod_{k=1}^{K} \left[\int_0^{F_k} dV_k \exp\left(i\omega f_k \left(1 - \frac{V_k}{F_k} \right) \right) + \int_{F_k}^{\infty} dV_k \right] \\ \times \frac{\sqrt{N}}{2\sigma_k V_k \sqrt{\pi z T}} \exp\left(- \frac{N(\ln(V_k/V_{k,0}) - (\mu - \sigma^2/2)T)^2}{4zT\sigma_k^2} \right) . \tag{4.71}$$

We rearranged the constants so that each term in $R(\omega, z)$ is normalized to unity. $R(\omega, z)$ can now be written as

$$R(\omega, z) = \exp\left(\sum_{k=1}^{K} \ln \left[\left(\int_0^{F_k} dV_k \overbrace{\exp\left(i\omega f_k \left(1 - \frac{V_k}{F_k} \right) \right)}^{Q(\omega, f_k)} + \int_{F_k}^{\infty} dV_k \right) \right. \right. \\ \left. \left. \times \frac{\sqrt{N}}{2\sigma_k V_k \sqrt{\pi z T}} \exp\left(- \frac{N(\ln(V_k/V_{k,0}) - (\mu - \sigma^2/2)T)^2}{4zT\sigma_k^2} \right) \right] \right) . \tag{4.72}$$

We write $Q(\omega, f_k)$ as

$$Q(\omega, f_k) = \sum_{m=0}^{\infty} \frac{(i\omega f_k)^m}{m!} \left(1 - \frac{V_k}{F_k} \right)^m . \tag{4.73}$$

Due to the normalization of $\langle p(\vec{V}) \rangle$, after insertion into Eq. (4.72) the non-default term and the integral over first term of $Q(\omega, f_k)$ become one. Thus, we can start the sum at $m = 1$ and obtain

$$R(\omega, z) = \exp\left(\sum_{k=1}^{K} \ln \left(1 + \sum_{m=1}^{\infty} \frac{(i\omega f_k)^m}{m!} M_{m,k}(z) \right) \right) \tag{4.74}$$

with

$$M_{m,k}(z) = \frac{\sqrt{N}}{2\sigma_k \sqrt{\pi z T}} \int_0^{F_k} \frac{1}{V_k} \left(1 - \frac{V_k}{F_k} \right)^m \\ \times \exp\left(- \frac{N(\ln(V_k/V_{k,0}) - (\mu - \sigma^2/2)T)^2}{4zT\sigma_k^2} \right) dV_k . \tag{4.75}$$

The integrals in Eq. (4.75) can be expressed with the generalized hypergeometrical function ${}_p\mathcal{F}_q$. However, this integral representation might be more intuitive. Moreover, for explicit $m = 1, 2$ the integrals can be calculated in a closed form using *Mathematica* [149], although this results in bulky expressions, as listed in appendix B.3. In the next step, we develop the logarithm in Eq. (4.74). Now we expand $\ln(1 + x)$ with

$$x \equiv \sum_{m=1}^{\infty} \frac{(i\omega f_k)^m}{m!} M_{m,k}(z) \tag{4.76}$$

as power series and obtain

$$\begin{aligned}\ln\left(1 + \sum_{m=1}^{3} \frac{(i\omega f_k)^m}{m!} M_{m,k}(z)\right) = &\sum_{m=1}^{3} \frac{(i\omega f_k)^m}{m!} M_{m,k}(z)\\ &-\frac{1}{2}\left(\sum_{m=1}^{3} \frac{(i\omega f_k)^m}{m!} M_{m,k}(z)\right)^2\\ &+\frac{1}{3}\left(\sum_{m=1}^{3} \frac{(i\omega f_k)^m}{m!} M_{m,k}(z)\right)^3\\ &+\dots\end{aligned} \tag{4.77}$$

Collecting all terms Eq. (4.77) up to the second order in f_k yields

$$\begin{aligned}\langle p(L)\rangle \approx &\frac{1}{2\pi\Gamma(N/2)} \int_0^{\infty} dz\, z^{\left(\frac{N}{2}-1\right)} \exp(-z) \int_{-\infty}^{+\infty} d\omega \exp(-i\omega L)\\ &\times \exp\left(\sum_{k=1}^{K}\left(i\omega M_{1,k}(z) f_k - \frac{\omega^2 f_k^2}{2}(M_{2,k}(z) - M_{1,k}(z)^2)\right)\right)\end{aligned} \tag{4.78}$$

$$\begin{aligned}= &\frac{1}{2\pi\Gamma(N/2)} \int_0^{\infty} dz\, z^{\left(\frac{N}{2}-1\right)} \exp(-z) \int_{-\infty}^{+\infty} d\omega\\ &\times \exp\left(i\omega\left(\left[\sum_{k=1}^{K} f_k M_{1,k}(z)\right] - L\right)\right.\\ &\left.-\frac{\omega^2}{2}\left[\sum_{k=1}^{K} f_k^2\left(M_{2,k}(z) - M_{1,k}(z)^2\right)\right]\right) .\end{aligned} \tag{4.79}$$

However, the convergence radius of the power series expansion involved in this approximation is one. Although we consider large portfolios K, i.e., f_k is small, ω runs from $-\infty$ to $+\infty$. This second-order approximation might describe the default terms adequately. However, the non-default terms, corresponding to a delta peak at $L = 0$ require ω to run from $-\infty$ to $+\infty$. Thus, the non-default terms cannot be approximated using this second-order approximation. To address this problem, we will develop an improved approximation in section 4.3.6. Now we solve the $d\omega$ integral in Eq. (4.79),

$$\langle p(L)\rangle \approx \frac{1}{\sqrt{2\pi}\Gamma(N/2)} \int_0^\infty dz\; z^{\left(\frac{N}{2}-1\right)} \exp(-z) \frac{1}{\sqrt{\widehat{M_2}(z)}} \exp\left(-\frac{(L-\widehat{M_1}(z))^2}{2\widehat{M_2}(z)}\right) \tag{4.80}$$

with

$$\widehat{M_1}(z) = \sum_{k=1}^{K} f_k M_{1,k}(z) \tag{4.81}$$

$$\widehat{M_2}(z) = \sum_{k=1}^{K} f_k^2 (M_{2,k}(z) - M_{1,k}(z)^2)\;. \tag{4.82}$$

However, due to the complexity of $M_1(z)$ and $M_2(z)$, the dz integral needs to be evaluated numerically. We will present this for an example of a homogeneous portfolio in the next section.

4.3.6 Homogeneous Portfolios

In case of a homogeneous portfolio, all credits have the same face value F, the same variance σ^2 and the same weight

$$f_k = \frac{1}{K}\;. \tag{4.83}$$

As $M_{1,k}(z)$ and $M_{1,k}(z)$ become identical for every k, we denote them with $M_1(z)$ and $M_1(z)$ leading to

$$\widehat{M}_1(z) = M_1(z) \tag{4.84}$$

$$\widehat{M}_2(z) = \frac{1}{K}(M_2(z) - M_1(z)^2) \tag{4.85}$$

$$M_m(z) = \frac{\sqrt{N}}{2\sigma\sqrt{\pi z T}} \int_0^F \frac{1}{V}\left(1 - \frac{V}{F}\right)^m \times \exp\left(-\frac{N(\ln(V/V_0) - (\mu - \sigma^2/2)T)^2}{4zT\sigma^2}\right) dV \,. \tag{4.86}$$

With insertion into Eq. (4.80), we can calculate the loss distribution for a homogeneous portfolio in the second order approximation. However, the dz integral needs to be evaluated numerically.

4.3.6.1 Improved Approximation

The second order approach can be improved by approximating the individual terms of the loss distribution instead of approximating the expression as a whole, similar as discussed in [31]. In case of a homogeneous portfolio, the combinatorial sum in Eq. (4.69) reduces to

$$\langle p(L)\rangle = \frac{1}{2\pi\Gamma(N/2)} \int_0^\infty dz\, z^{\left(\frac{N}{2}-1\right)} \exp(-z) \int_{-\infty}^{+\infty} d\omega \exp(-i\omega L) \times \sum_{j=0}^{K} \binom{K}{j} \left(R^{(\mathrm{D})}(\omega, z)\right)^j \left(R^{(\mathrm{ND})}(z)\right)^{K-j} \,, \tag{4.87}$$

with the non-default term

$$\left(R^{(\mathrm{ND})}\right)^{K-j} = \left(\int_F^\infty dV \frac{\sqrt{N}}{2\sigma V\sqrt{\pi z T}} \times \exp\left(-\frac{N(\ln(V/V_0) - (\mu - \sigma^2/2)T)^2}{4zT\sigma^2}\right)\right)^{K-j} \tag{4.88}$$

$$= \left(\frac{1}{2} + \frac{1}{2}\mathrm{Erf}\left[\frac{\sqrt{N}(\ln(F/V_0) - (\mu - \sigma^2/2)T)}{2\sigma\sqrt{zT}}\right]\right)^{K-j} \tag{4.89}$$

and the default term

$$\left(R^{(\mathrm{D})}(\omega,z)\right)^j = \Bigg(\int_0^F dV \overbrace{\exp\left(\frac{i\omega}{K}\left(1-\frac{V}{F}\right)\right)}^{Q(\omega,K)} \times \frac{\sqrt{N}}{2\sigma V\sqrt{\pi z T}} \exp\left(-\frac{N(\ln(V/V_0)-(\mu-\sigma^2/2)T)^2}{4zT\sigma^2}\right)\Bigg)^j . \tag{4.90}$$

Analogously to the previous section, we develop $Q(\omega, K)$ and consequently develop the logarithm, resulting in

$$\int_{-\infty}^{+\infty} d\omega \exp(-i\omega L) \left(R^{(\mathrm{D})}(\omega,z)\right)^j$$

$$= \int_{-\infty}^{+\infty} d\omega \exp\left(i\omega\left(\frac{j}{K}M_1(z)-L\right)-\frac{\omega^2 j}{2K^2}\left(M_2(z)-M_1(z)^2\right)\right) \tag{4.91}$$

$$= \sqrt{\frac{2\pi K^2}{j\left(M_2(z)-M_1(z)^2\right)}} \exp\left(-\frac{(LK-jM_1(z))^2}{2j\left(M_2(z)-M_1(z)^2\right)}\right) . \tag{4.92}$$

In this approximation, the non-default terms given by Eq. (4.89), can be calculated exactly. They correspond to a delta peak at $L = 0$. Another advantage over the approximation presented in Eq. (4.80) is that the approximation is performed for each number of defaults j separately and weighted by j/K accordingly. In this approximation, the omitted third term is of the order j/K^3 and thereby much smaller than the third term of the simple second order approximation (4.86), which would be of the order $1/K^2$. Thus, when approximating each term in the combinatorial sum separately,

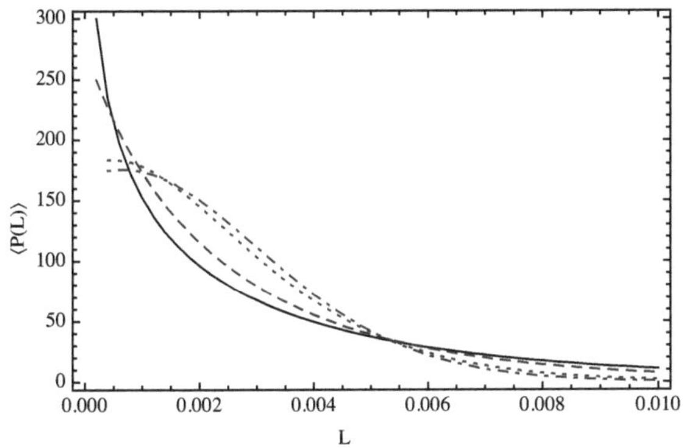

Figure 4.5: The loss distribution for $K = 10$, $\sigma = 0.15$, $\mu = 0.05$, $T = 1$, $V_0 = 100$, $F = 75$ and different strengths of correlations N: $N = K$ (solid), $N = 2K$ (dashed), $N = 10K$ (dotted), $N = 30K$ (dot-dashed).

we obtain an improved approximation. Insertion into (4.87) leads to

$$\langle p(L)\rangle \approx \frac{1}{2\pi\Gamma(N/2)} \sum_{j=0}^{K} \binom{K}{j} \int_0^\infty dz\; z^{\left(\frac{N}{2}-1\right)} \exp\left(-z\right)$$
$$\left(\sqrt{\frac{2\pi K^2}{j\left(M_2(z) - M_1(z)^2\right)}} \exp\left(-\frac{(LK - jM_1(z))^2}{2j\left(M_2(z) - M_1(z)^2\right)}\right)\right.$$
$$\left.\times \left(\frac{1}{2} + \frac{1}{2}\mathrm{Erf}\left[\frac{\ln(F/V_0) - (\mu - \sigma^2/2)T}{2\sigma\sqrt{zT}}\right]\right)^{K-j}\right), \tag{4.93}$$

which is the final result.

4.4 Application

We now apply the analytically developed model to a specific example. To analyze the impact of correlations, we calculate the loss distribution for different homogeneous portfolios with sizes $K = 10$, $K = 50$ and $K = 100$ with the parameters $V_0 = 100$, $\mu = 0.05$, $\sigma = 0.15$, $F = 75$ and $T = 1$. As stated in the previous section, we can control the amount of correlation in our model with the parameter N. $N = K$ corresponds to the strongest impact of random correlations. For $N \to \infty$, the correlation matrix becomes

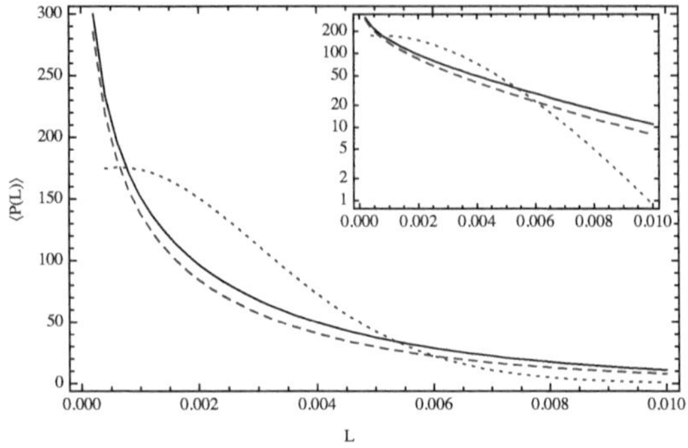

(a) K=10

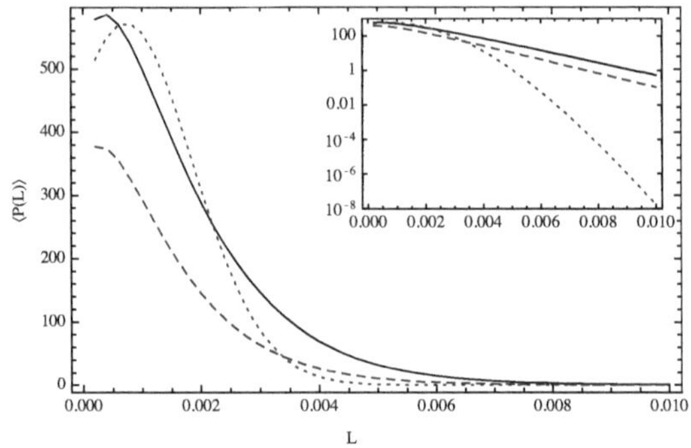

(b) K=50

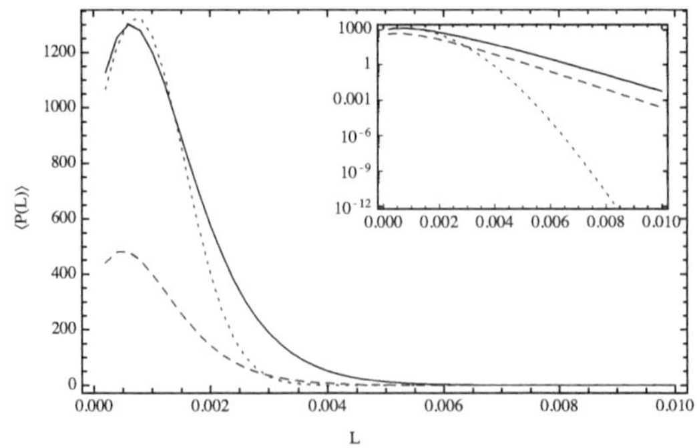

(c) K=100

Figure 4.6: The loss distribution of a homogeneous portfolio with $\sigma = 0.15$, $\mu = 0.05$, $T = 1$, $V_0 = 100$, $F = 75$ and different values of K. The dashed line represents the simple approximation; the solid line represents the improved approximation. Both have been calculated with maximum random correlations, $N = K$. The uncorrelated case is given by the dotted line, calculated with the improved approximation with $N = 30K$.

the identity matrix. Thus, this represents the transition to a system without correlations. As we have to evaluate the loss distributions numerically, $N \to \infty$ has to be properly interpetrated. Hence we need to identify a value for which this convergence is valid in good approximation.

Fig. 4.5 illustrates exemplarily the loss distribution for $K = 10$ and different values of N. Our study indicates that a value of $N = 30K$ is a good choice for approximating the uncorrelated case and is still numerically feasible.

The results are presented in Fig. 4.6. For all portfolio sizes, $K = 10$, $K = 50$ and $K = 100$, we obtain heavier tails of the loss distribution of the correlated portfolio compared to the uncorrelated case. Even the second order approximation, represented by the dashed curve, exhibits these heavy tails. Using the inserted logarithmic plots, we can identify a nearly power-law decay of the loss distribution for the correlated case.

The distributions become narrower for larger values of K. However, the tails of the correlated case remain heavier than those of the uncorrelated case. While for $K = 10$, both approximations give similar results, their difference becomes larger with K. As both approximations have to be performed numerically, we suggest to always use the improved approximation. However, the tail behavior remains the same, even for the second order approximation, as indicated by the logarithmic scaled inserts in Fig. 4.6. This is a strong indication that the tails of the loss distribution are vastly underestimated, if correlations are not taken into account.

Due to the approximation, the normalization of the loss distribution is not exact. Especially the normalization of the second order approximation is poor for large values of K. The normalization might also be used as an indication for the quality of the approximation. The improved approximation exhibits a delta peak at $L = 0$, as the non-default terms can be calculated exactly. However, the interval $[0; 0.0002[$ was not evaluated due to numerical feasibility.

In our example, we do not vary the maturity time T, i.e., we choose $T = 1$. If the model is properly set up, one can increase T to estimate the evolution of the loss distribution. However, this evolution depends strongly on the drifts μ_k and standard deviations σ_k. Depending on their value, the exposure to default risk can either increase or decrease.

Variable	Description	Unit
K	Number of assets	–
T	Time of maturity	[year]
σ_k	volatility of the k-th asset	$[\text{year}]^{-1/2}$
μ_k	drift of the k-th asset	$[\text{year}]^{-1}$
N	Parameter to control correlations, $N \to \infty$: uncorrelated limit	–
$V_{k,0}$	start price of the k-th asset	[currency]
F_k	face value of the k-th asset	[currency]

Table 4.1: Input of the structural credit risk model.

4.5 Summary

Our results clearly demonstrate that the risk in a credit portfolio is heavily underestimated if correlations are not taken into account in the quantification of risk. Even with random correlations and an average correlation level of zero, we obtain distributions whose tails differ in several orders of magnitude. In contrast, the probability of large losses in uncorrelated portfolios is significantly reduced within the model. This demonstrates that the effect of diversification, i.e., the reduction of risk in large portfolios, does not give good results in credit portfolios. Even the presence of random correlations around zero lowers the effect of diversification dramatically.

Our model can be used to estimate the lower bound of the risk embedded in a credit portfolio. The eigenvalue density of empirical correlation matrices show that the amount of randomness in stock market return correlations is considerable [11]. Our model is directly applicable to speculative margin loans, i.e., credits that are based on an investor's stock portfolio. An overview of the model's input parameters is given in Tab. 4.1.

In our model, some features are not taken into account which are present in empirical data, such as jumps or an overall positive correlation level. Those features are difficult to treat completely analytically. However, even in our simple setup we obtain a heavy-tailed loss distribution

The results are especially relevant for *CDOs (Collateralized Debt Obligations)*, bundles of credits that are traded on equity markets. CDOs are constructed in order to lower the overall risk. The components of a CDO can be exposed to large risks. It is often believed that the CDO has a significantly lower risk. We showed that this diversification only works well if the correlations in the credit portfolio are identical to zero. Even if the average correlation level is zero, but the individual correlations fluctuate, we obtain a heavy-tailed loss distribution. However, the average correlation level, for example, of stock returns is usually positive, corresponding to the market risk. A common approach is to remove this kind of risk by hedging. This corresponds to a shift in the correlation matrix by the size of the market correlation. The result is a correlation matrix with average correlation level of zero, as we assumed in our model. Thus, even if a portfolio manager eliminates the market risk, the probability of large losses is still significant. In credit portfolios, the risk of simultaneous defaults is not only caused by positive correlations, but also by their fluctuations.

5 Conclusions

Important aspects of financial markets were studied as complex systems. The common feature of the studies is the estimation and identification of statistical dependencies. In this matter, concepts of statistical physics were used, such as diffusion processes or Random Matrix Theory, to gain deeper insight into the mechanisms of financial markets. Moreover, the traditional strength of theoretical physics, the mathematical modeling, was employed to develop methods that on the one hand improve the estimation of statistical dependence significantly, and on the other hand permit new insight into the dynamics of the dependence structure.

The latter was provided by the development of a similarity measure. By reducing central aspects of the statistical dependence of a financial market to a simple representation, this allows one to study and visualize the general evolution of the financial market. Disclosing the dynamics is a key challenge, as financial markets show non-stationarity behavior. As a large financial market can be seen as representative for a whole economy, this can permit to identify tensions in the market before their full emergence to a financial crisis, for example by recognizing similarities between market crashes in the past. Moreover, the similarity measure was used to identify typical states between which the financial market switches back and forth.

One important application of this measure is risk management. By providing a simple indicatior for an economy's current status, a portfolio manager can react timely or learn lessons from similar crises in the past. This application was demonstrated in an empirical study, where it led to a significant reduction of risk. However, due to the elementary character of this method, it can possibly be applied to a wide range of other complex systems. In several complex systems, it is possible to obtain a large number of correlated data over time. Such systems include, but are not restricted to, biological or medical time series, such as EEG, chemical and nuclear reactions or weather data. All systems share the common attribute, that their non-stationarity behavior is of particular importance when it comes to abnormal events.

The measurement of correlations is a powerful instrument to reduce the complexity of a system in order to treat it analytically. However, correlation coefficients also have major disadvantages. By reducing the statistical dependence to a single number, they imply a linear statistical dependence although the dependence structure of stock returns is usually non-linear. This is especially true for rare events that have a large impact, i.e., large negative or positive returns. Using copulae one can describe the statistical dependence much more precisely. Moreover, because copulae represent a scale-free measure, it is possible to compare the dependence of different systems, even if their individual marginal distributions have different shapes.

To disclose the degree of error involved in a correlation coefficient, the average pairwise copula of the U.S. stock market was estimated in a large-scale empirical study. By comparing the results to the Gaussian copula, the copula that is implied by a Pearson correlation coefficient, differences could be quantified. Furthermore, key features of the empirical copula were mapped to the corresponding correlation coefficient. As correlation coefficients are convenient to use and common throughout the whole financial industry, the disclosed relation allows correcting the estimation of risk in existing applications.

Another approach to improve models that include correlation estimates was considered in the study of the Epps effect. The Epps effect describes the decline of correlation estimates in high frequency data. Since the phenomenon was discovered in 1979, many models have been developed to compensate for this effect. Almost all of these models share the common requirement for calibration and many of these models are incompatible with each other. A new approach was carried out in this work. Rather than aiming for a complete compensation of the Epps effect, causes were identified that contribute to the effect on a completely statistical basis. Hence, compensation methods for these causes could be developed and verified in a model setup. These compensation methods do not require model calibrations or fitting parameters. In an empirical study, it was demonstrated that these statistical causes can be responsible for a large portion of up to 75% of the Epps effect. By including these causes into existing practices, it is possible to estimate correlations more precisely and thus to estimate the risk more realistically. Just like the aforementioned similarity measure, this method is not restricted to financial data. It can be applied to any scenario where correlations of asynchronous or discretized time series are estimated.

A further focus of this work was devoted to credit risk. This estimation of credit risk differs significantly from other kinds of financial risk. This is due to the complex nature of a credit process. The exposure to the risk of default, the case where a credit is not paid back completely, leads to a highly asymmetric probability distribution of losses. A concept of statistical physics, Random Matrix Theory, was used to estimate the average loss distribution. By averaging over all possible combinations of correlations, it is demonstrated that the presence of correlations severely limits the effect of diversification. This is true even though the average correlation level is zero. The exposure to risk in a credit portfolio is vastly underestimated, if correlations are not taken into account. The results can be seen as lower bound for the estimation of credit risk.

All topics covered in this work have in common that failures in these fields were a key factor in the emergence of the financial crisis of 2008–2009. The developed methods can help recognizing a future financial crisis in its beginning. The improved estimation of risk aims at preventing future financial crises at their source. Certainly, the models can only capture quantitative aspects of an economy. For example, political or physiological factors can only be accounted for indirectly. However, quantitative models represent a powerful and mandatory tool to estimate risk in global financial markets that become increasingly complex.

A Stock Ensembles

GICS Branch	Stock 1				Stock 2				corr	var_{corr}
	Symbol	Name	Stock Exchange	Volume	Symbol	Name	Stock Exchange	Volume		
Consumer Discretionary	AMZN	Amazon Corp.	NASDAQ	1215700	SBUX	Starbucks Corp.	NASDAQ	1287200	0.80	0.45
	APOL	Apollo Group	NASDAQ	332200	SHLD	Sears Holdings Corporation	NASDAQ	317900	0.28	0.57
	AMZN	Amazon Corp.	NASDAQ	1215700	SPLS	Staples Inc.	NASDAQ	674500	0.80	0.62
	AMZN	Amazon Corp.	NASDAQ	1215700	CMCSA	Comcast Corp.	NASDAQ	910000	0.69	0.64
	EXPE	Expedia Inc.	NASDAQ	883300	SHLD	Sears Holdings Corporation	NASDAQ	317900	0.43	0.66
Consumer Staples	COST	Costco Co.	NASDAQ	237500	PG	Procter & Gamble	NYSE	1396077700	0.05	0.71
	COST	Costco Co.	NASDAQ	237500	CVS	CVS Caremark Corp.	NYSE	1454768200	0.09	0.77
	KO	Coca Cola Co.	NYSE	1152559200	COST	Costco Co.	NASDAQ	237500	0.04	0.79
	MO	Altria Group Inc.	NYSE	1312801100	CCE	Coca-Cola Enterprises	NYSE	335449400	0.29	0.79
	CCE	Coca-Cola Enterprises	NYSE	335449400	KFT	Kraft Foods Inc-A	NYSE	1438368700	0.32	0.80
Energy	EP	El Paso Corp.	NYSE	697103700	SE	Spectra Energy Corp.	NYSE	331523100	0.34	0.84
	CVX	Chevron Corp.	NYSE	1271849400	SE	Spectra Energy Corp.	NYSE	331523100	0.26	0.85
	HES	Hess Corporation	NYSE	427802700	SE	Spectra Energy Corp.	NYSE	331523100	0.23	0.85
	MUR	Murphy Oil	NYSE	223916300	SE	Spectra Energy Corp.	NYSE	331523100	0.25	0.86
	SII	Smith International	NYSE	370893400	SE	Spectra Energy Corp.	NYSE	331523100	0.32	0.86
Financials	SCHW	Charles Schwab	NASDAQ	1445500	ETFC	E*Trade Financial Corp.	NASDAQ	1391800	0.67	0.44
	ETFC	E*Trade Financial Corp.	NASDAQ	1391800	FITB	Fifth Third Bancorp	NASDAQ	218000	0.29	0.58
	SCHW	Charles Schwab	NASDAQ	1445500	FITB	Fifth Third Bancorp	NASDAQ	218000	0.36	0.58
	SCHW	Charles Schwab	NASDAQ	1445500	HCBK	Hudson City Bancorp	NASDAQ	686900	0.54	0.60
	ACAS	American Capital Strategies Ltd	NASDAQ	207200	SCHW	Charles Schwab	NASDAQ	1445500	0.50	0.60
Health Care	CELG	Celgene Corp.	NASDAQ	619200	ESRX	Express Scripts	NASDAQ	998400	0.65	0.47
	AMGN	Amgen	NASDAQ	813900	CELG	Celgene Corp.	NASDAQ	619200	0.48	0.50
	AMGN	Amgen	NASDAQ	813900	BIIB	BIOGEN IDEC Inc.	NASDAQ	381800	0.48	0.52
	CELG	Celgene Corp.	NASDAQ	619200	THC	Tenet Healthcare Corp.	NYSE	805228900	-0.23	0.53
	AMGN	Amgen	NASDAQ	813900	GENZ	Genzyme Corp.	NASDAQ	242900	0.57	0.53
Industrials	GE	General Electric	NYSE	4303823300	LUV	Southwest Airlines	NYSE	862775700	0.53	0.73
	MMM	3M Company	NYSE	549124400	CBE	Cooper Industries Ltd.	NYSE	175911300	0.24	0.73
	CBE	Cooper Industries Ltd.	NYSE	175911300	GWW	Grainger (W.W.) Inc.	NYSE	95324400	0.15	0.73
	CBE	Cooper Industries Ltd.	NYSE	175911300	GR	Goodrich Corporation	NYSE	144177800	0.15	0.75
	CBE	Cooper Industries Ltd.	NYSE	175911300	FLR	Fluor Corp. (New)	NYSE	171713100	0.10	0.76
Information Technology	AAPL	Apple Inc.	NASDAQ	4627500	INTC	Intel Corp.	NASDAQ	5529100	0.84	0.19
	AAPL	Apple Inc.	NASDAQ	4627500	CSCO	Cisco Systems	NASDAQ	4886800	0.76	0.24
	AAPL	Apple Inc.	NASDAQ	4627500	YHOO	Yahoo Inc.	NASDAQ	2609700	0.71	0.25
	AAPL	Apple Inc.	NASDAQ	4627500	ORCL	Oracle Corp.	NASDAQ	1731900	0.78	0.25
	AAPL	Apple Inc.	NASDAQ	4627500	EBAY	eBay Inc.	NASDAQ	1056100	0.73	0.26
Materials	MON	Monsanto Co.	NYSE	479605700	SEE	Sealed Air Corp.(New)	NYSE	126716200	0.06	0.52
	FCX	Freeport-McMoran Cp & Gld	NYSE	1058215000	SIAL	Sigma-Aldrich	NASDAQ	133800	0.05	0.54
	ECL	Ecolab Inc.	NYSE	163404500	SEE	Sealed Air Corp.(New)	NYSE	126716200	0.26	0.60
	ATI	Allegheny Technologies Inc	NYSE	269746100	SEE	Sealed Air Corp.(New)	NYSE	126716200	0.10	0.63
	PX	Praxair Inc.	NYSE	245761900	SEE	Sealed Air Corp.(New)	NYSE	126716200	0.12	0.65
Telecommunication Services	Q	Qwest Communications Int	NYSE	1623807700	S	Sprint Nextel Corp.	NYSE	2044634000	0.52	0.84
	Q	Qwest Communications Int	NYSE	1623807700	VZ	Verizon Communications	NYSE	1472335800	0.49	0.86
	S	Sprint Nextel Corp.	NYSE	2044634000	VZ	Verizon Communications	NYSE	1472335800	0.49	0.87
	AMT	American Tower Corp.	NYSE	387199400	Q	Qwest Communications Int	NYSE	1623807700	0.26	0.97
	AMT	American Tower Corp.	NYSE	387199400	WIN	Windstream Corporation	NYSE	400634200	0.10	0.99
Utilities	DUK	Duke Energy	NYSE	902519200	DYN	Dynegy Inc.	NYSE	702035600	0.54	0.74
	CMS	CMS Energy	NYSE	264225200	DYN	Dynegy Inc.	NYSE	702035600	0.48	0.79
	CMS	CMS Energy	NYSE	264225200	DUK	Duke Energy	NYSE	902519200	0.39	0.81
	AES	AES Corp.	NYSE	556049300	CMS	CMS Energy	NYSE	264225200	0.31	0.84
	CNP	CenterPoint Energy	NYSE	359757800	DUK	Duke Energy	NYSE	902519200	0.33	0.84

Table A.1: Top 5 five stock pairs with the most stable correlation from each GICS branch of the S&P 500 index.

GICS Branch	Stock 1				Stock 2				corr	var$_{corr}$
	Symbol	Name	Stock Exchange	Volume	Symbol	Name	Stock Exchange	Volume		
Consumer Discretionary	APOL	Apollo Group	NASDAQ	332200	SPLS	Staples Inc.	NASDAQ	674500	0.77	1.08
	BBBY	Bed Bath & Beyond	NASDAQ	233000	SPLS	Staples Inc.	NASDAQ	674500	0.79	1.08
	AMZN	Amazon Corp.	NASDAQ	1215700	SPLS	Staples Inc.	NASDAQ	674500	0.80	0.62
	AMZN	Amazon Corp.	NASDAQ	1215700	SBUX	Starbucks Corp.	NASDAQ	1287200	0.80	0.45
	SPLS	Staples Inc.	NASDAQ	674500	SBUX	Starbucks Corp.	NASDAQ	1287200	0.81	0.69
Consumer Staples	KO	Coca Cola Co.	NYSE	1152559200	SLE	Sara Lee Corp.	NYSE	542934700	0.42	0.90
	SLE	Sara Lee Corp.	NYSE	542934700	WMT	Wal-Mart Stores	NYSE	1992433400	0.43	1.02
	CVS	CVS Caremark Corp.	NYSE	1454768200	KFT	Kraft Foods Inc-A	NYSE	1438368700	0.44	0.98
	KFT	Kraft Foods Inc-A	NYSE	1438368700	SLE	Sara Lee Corp.	NYSE	542934700	0.48	0.87
	COST	Costco Co.	NASDAQ	237500	WFMI	Whole Foods Market	NASDAQ	211400	0.59	1.09
Energy	EP	El Paso Corp.	NYSE	697103700	RDC	Rowan Cos.	NYSE	369346200	0.39	0.91
	CVX	Chevron Corp.	NYSE	1271849400	XOM	Exxon Mobil Corp.	NYSE	2798325600	0.41	1.00
	XOM	Exxon Mobil Corp.	NYSE	2798325600	HAL	Halliburton Co.	NYSE	1701703200	0.45	1.18
	CVX	Chevron Corp.	NYSE	1271849400	EP	El Paso Corp.	NYSE	697103700	0.45	0.88
	COP	ConocoPhillips	NYSE	1382115600	EP	El Paso Corp.	NYSE	697103700	0.46	0.94
Financials	CINF	Cincinnati Financial	NASDAQ	55900	TROW	T. Rowe Price Group	NASDAQ	216400	0.60	1.74
	FITB	Fifth Third Bancorp	NASDAQ	218000	HBAN	Huntington Bancshares	NASDAQ	71600	0.62	2.01
	FITB	Fifth Third Bancorp	NASDAQ	218000	TROW	T. Rowe Price Group	NASDAQ	216400	0.63	1.34
	SCHW	Charles Schwab	NASDAQ	1445500	ETFC	E*Trade Financial Corp.	NASDAQ	1391800	0.67	0.44
	TROW	T. Rowe Price Group	NASDAQ	216400	ZION	Zions Bancorp	NASDAQ	48500	0.70	1.40
Health Care	BIIB	BIOGEN IDEC Inc.	NASDAQ	381800	PDCO	Patterson Cos. Inc.	NASDAQ	106400	0.60	1.15
	BIIB	BIOGEN IDEC Inc.	NASDAQ	381800	GENZ	Genzyme Corp.	NASDAQ	242900	0.62	0.85
	GENZ	Genzyme Corp.	NASDAQ	242900	GILD	Gilead Sciences	NASDAQ	275100	0.64	0.91
	CELG	Celgene Corp.	NASDAQ	619200	ESRX	Express Scripts	NASDAQ	998400	0.65	0.47
	BSX	Boston Scientific	NYSE	1205569800	THC	Tenet Healthcare Corp.	NYSE	805228900	0.68	0.60
Industrials	GE	General Electric	NYSE	4303823300	LUV	Southwest Airlines	NYSE	862775700	0.53	0.73
	CTAS	Cintas Corporation	NASDAQ	48500	EXPD	Expeditors Int'l	NASDAQ	215600	0.54	1.48
	EXPD	Expeditors Int'l	NASDAQ	215600	MNST	Monster Worldwide	NASDAQ	196700	0.55	1.20
	CHRW	C.H. Robinson Worldwide	NASDAQ	112400	EXPD	Expeditors Int'l	NASDAQ	215600	0.56	1.46
	MNST	Monster Worldwide	NASDAQ	196700	PCAR	PACCAR Inc.	NASDAQ	164400	0.56	1.23
Information Technology	DELL	Dell Inc.	NASDAQ	909400	ORCL	Oracle Corp.	NASDAQ	1731900	0.79	0.47
	CSCO	Cisco Systems	NASDAQ	4886800	DELL	Dell Inc.	NASDAQ	909400	0.79	0.41
	INTC	Intel Corp.	NASDAQ	5529100	ORCL	Oracle Corp.	NASDAQ	1731900	0.82	0.29
	CSCO	Cisco Systems	NASDAQ	4886800	ORCL	Oracle Corp.	NASDAQ	1731900	0.83	0.34
	AAPL	Apple Inc.	NASDAQ	4627500	INTC	Intel Corp.	NASDAQ	5529100	0.84	0.19
Materials	DD	Du Pont (E.I.)	NYSE	716944300	FCX	Freeport-McMoran Cp & Gld	NYSE	1058215000	0.36	1.02
	FCX	Freeport-McMoran Cp & Gld	NYSE	1058215000	MON	Monsanto Co.	NYSE	479605700	0.36	0.85
	APD	Air Products & Chemicals	NYSE	187953500	BLL	Ball Corp.	NYSE	119560600	0.36	0.88
	ECL	Ecolab Inc.	NYSE	163404500	NEM	Newmont Mining Corp. (Hldg. Co.)	NYSE	958900000	0.37	0.93
	FCX	Freeport-McMoran Cp & Gld	NYSE	1058215000	NEM	Newmont Mining Corp. (Hldg. Co.)	NYSE	958900000	0.43	1.06
Telecommunication Services	T	AT&T Inc.	NYSE	2663617200	Q	Qwest Communications Int	NYSE	1623807700	0.45	1.01
	S	Sprint Nextel Corp.	NYSE	2044634000	VZ	Verizon Communications	NYSE	1472335800	0.49	0.87
	Q	Qwest Communications Int	NYSE	1623807700	VZ	Verizon Communications	NYSE	1472335800	0.49	0.86
	T	AT&T Inc.	NYSE	2663617200	VZ	Verizon Communications	NYSE	1472335800	0.50	1.05
	Q	Qwest Communications Int	NYSE	1623807700	S	Sprint Nextel Corp.	NYSE	2044634000	0.52	0.84
Utilities	DUK	Duke Energy	NYSE	902519200	TE	TECO Energy	NYSE	177983100	0.35	0.87
	D	Dominion Resources	NYSE	321656100	XEL	Xcel Energy Inc	NYSE	337262500	0.36	0.92
	CMS	CMS Energy	NYSE	264225200	DUK	Duke Energy	NYSE	902519200	0.39	0.81
	CMS	CMS Energy	NYSE	264225200	DYN	Dynegy Inc.	NYSE	702035600	0.48	0.79
	DUK	Duke Energy	NYSE	902519200	DYN	Dynegy Inc.	NYSE	702035600	0.54	0.74

Table A.2: Top 5 five stock pairs with the highest correlation from each GICS branch of the S&P 500 index.

Symbol	Name	Stock ex.	$\langle S\rangle$	σ_S	$\sigma_S/\langle S\rangle$	Avrg. trad.
BSC	Bear Stearns Cos.	NYSE	133.14	23.90	0.180	25529
SII	Smith International	NYSE	57.12	10.46	0.183	13949
FRE	Federal Home Loan Mtg.	NYSE	57.87	10.80	0.187	21389
LSI	LSI Corporation	NYSE	7.93	1.49	0.187	20497
PCAR	PACCAR Inc.	NASDAQ	74.10	13.89	0.187	11376
JCP	Penney (J.C.)	NYSE	69.49	13.08	0.188	15925
CVG	Convergys Corp.	NYSE	21.77	4.16	0.191	5147
CBG	CB Richard Ellis Group	NYSE	31.58	6.08	0.193	13992
LIZ	Liz Claiborne Inc.	NYSE	35.80	6.97	0.195	6330
BC	Brunswick Corp.	NYSE	28.09	5.54	0.197	5387
CNX	CONSOL Energy Inc.	NYSE	45.20	8.99	0.199	12842
AKAM	Akamai Technologies Inc	NASDAQ	42.89	8.63	0.201	22620
MCO	Moody's Corp	NYSE	56.88	11.87	0.209	16294
RSH	RadioShack Corp	NYSE	24.72	5.17	0.209	14454
LUK	Leucadia National Corp.	NYSE	38.25	8.05	0.211	3429
FLR	Fluor Corp. (New)	NYSE	115.34	24.90	0.216	7413
NCC	National City Corp.	NYSE	30.54	6.71	0.220	17876
LXK	Lexmark Int'l Inc	NYSE	48.47	10.91	0.225	8838
DF	Dean Foods	NYSE	32.98	7.49	0.227	6647
MBI	MBIA Inc.	NYSE	58.90	13.50	0.229	17571
FCX	Freeport-McMoran Cp & Gld	NYSE	82.68	19.07	0.231	45292
FHN	First Horizon National	NYSE	34.09	7.88	0.231	7319
ESRX	Express Scripts	NASDAQ	70.84	16.43	0.232	11623
JNPR	Juniper Networks	NASDAQ	27.11	6.35	0.234	33006
DDS	Dillard Inc.	NYSE	28.97	6.79	0.234	8291
CMI	Cummins Inc.	NYSE	117.32	47.60	0.406	9909
JNY	Jones Apparel Group	NYSE	26.09	6.24	0.239	5803
MON	Monsanto Co.	NYSE	70.65	16.94	0.240	17403
SOV	Sovereign Bancorp	NYSE	20.13	4.84	0.240	10837
CMCSA	Comcast Corp.	NASDAQ	27.05	6.64	0.246	55284
OMX	OfficeMax Inc.	NYSE	39.68	9.85	0.248	6009
WM	Washington Mutual	NYSE	36.55	9.09	0.249	39145
KG	King Pharmaceuticals	NYSE	16.29	4.15	0.254	9548
JEC	Jacobs Engineering Group	NYSE	71.33	32.20	0.451	5501
CIT	CIT Group	NYSE	46.43	12.31	0.265	11141
THC	Tenet Healthcare Corp.	NYSE	5.66	1.51	0.267	12112
KBH	KB Home	NYSE	37.21	10.34	0.278	16670
GME	GameStop Corp.	NYSE	47.04	23.36	0.497	9619
CTX	Centex Corp.	NYSE	37.79	10.53	0.279	14328
CTSH	Cognizant Technology Solutions	NASDAQ	72.31	20.33	0.281	13112
ODP	Office Depot	NYSE	28.13	7.92	0.282	14062
NOV	National Oilwell Varco Inc.	NYSE	88.30	30.40	0.344	19716
GILD	Gilead Sciences	NASDAQ	57.24	17.77	0.310	26240
ABK	Ambac Financial Group	NYSE	70.98	23.27	0.328	16261
PHM	Pulte Homes Inc.	NYSE	21.82	7.34	0.336	15737
LEN	Lennar Corp.	NYSE	35.41	11.91	0.336	16250
MTG	MGIC Investment	NYSE	46.61	17.01	0.365	14053
CC	Circuit City Group	NYSE	13.65	5.00	0.366	16660
ETFC	E*Trade Financial Corp.	NASDAQ	17.65	6.86	0.389	33380
CFC	Countrywide Financial Corp.	NYSE	28.75	11.41	0.397	65703

Table A.3: Ensemble of 50 stocks from the S&P 500 index. The stocks provide the highest relation between the mean price $\langle S\rangle$ and its standard deviation $\sigma_S/\langle S\rangle$ with at least 1000 trades per day.

Stock 1				Stock 2				corr	var_{corr}
F	Ford Motor	NYSE	8.16	Q	Qwest Communications Int	NYSE	8.63	0.11	0.02
Q	Qwest Communications Int	NYSE	8.63	CPWR	Compuware Corp.	NASDAQ	9.44	0.12	0.03
CPWR	Compuware Corp.	NASDAQ	9.44	UIS	Unisys Corp.	NYSE	7.65	0.13	0.10
CPWR	Compuware Corp.	NASDAQ	9.44	THC	Tenet Healthcare Corp.	NYSE	5.66	0.14	0.03
NOVL	Novell Inc.	NASDAQ	7.21	F	Ford Motor	NYSE	8.16	0.14	0.02
F	Ford Motor	NYSE	8.16	LSI	LSI Corporation	NYSE	7.93	0.15	0.03
Q	Qwest Communications Int	NYSE	8.63	THC	Tenet Healthcare Corp.	NYSE	5.66	0.15	0.03
F	Ford Motor	NYSE	8.16	CPWR	Compuware Corp.	NASDAQ	9.44	0.15	0.02
Q	Qwest Communications Int	NYSE	8.63	LSI	LSI Corporation	NYSE	7.93	0.16	0.03
UIS	Unisys Corp.	NYSE	7.65	THC	Tenet Healthcare Corp.	NYSE	5.66	0.17	0.05
Q	Qwest Communications Int	NYSE	8.63	UIS	Unisys Corp.	NYSE	7.65	0.18	0.05
DYN	Dynegy Inc.	NYSE	8.77	THC	Tenet Healthcare Corp.	NYSE	5.66	0.18	0.02
Q	Qwest Communications Int	NYSE	8.63	DYN	Dynegy Inc.	NYSE	8.77	0.19	0.03
DYN	Dynegy Inc.	NYSE	8.77	LSI	LSI Corporation	NYSE	7.93	0.21	0.03
UIS	Unisys Corp.	NYSE	7.65	LSI	LSI Corporation	NYSE	7.93	0.22	0.05
LSI	LSI Corporation	NYSE	7.93	THC	Tenet Healthcare Corp.	NYSE	5.66	0.22	0.02
NOVL	Novell Inc.	NASDAQ	7.21	UIS	Unisys Corp.	NYSE	7.65	0.22	0.04
CPWR	Compuware Corp.	NASDAQ	9.44	LSI	LSI Corporation	NYSE	7.93	0.23	0.04
F	Ford Motor	NYSE	8.16	UIS	Unisys Corp.	NYSE	7.65	0.23	0.04
DYN	Dynegy Inc.	NYSE	8.77	CPWR	Compuware Corp.	NASDAQ	9.44	0.24	0.03
NOVL	Novell Inc.	NASDAQ	7.21	CPWR	Compuware Corp.	NASDAQ	9.44	0.25	0.06
NOVL	Novell Inc.	NASDAQ	7.21	DYN	Dynegy Inc.	NYSE	8.77	0.26	0.04
NOVL	Novell Inc.	NASDAQ	7.21	LSI	LSI Corporation	NYSE	7.93	0.29	0.07
DYN	Dynegy Inc.	NYSE	8.77	UIS	Unisys Corp.	NYSE	7.65	0.29	0.03
F	Ford Motor	NYSE	8.16	DYN	Dynegy Inc.	NYSE	8.77	0.39	0.03

Table A.4: Ensemble of the 50 highest correlated stock pairs from the S&P 500 index that are averagely traded between USD 0.01 and USD 10.00. The column var_{corr} refers to the variance of the correlation of a moving 30-day window.

Stock 1				Stock 2				corr	var_{corr}
CMS	CMS Energy	NYSE	17.18	CZN	Citizens Communications	NYSE	14.34	0.37	0.05
XRX	Xerox Corp.	NYSE	17.45	AW	Allied Waste Industries	NYSE	12.71	0.37	0.04
MU	Micron Technology	NYSE	11.38	TER	Teradyne Inc.	NYSE	15.07	0.37	0.04
IPG	Interpublic Group	NYSE	11.23	HBAN	Huntington Bancshares	NASDAQ	19.95	0.38	0.03
TE	TECO Energy	NYSE	16.96	HBAN	Huntington Bancshares	NASDAQ	19.95	0.38	0.05
TE	TECO Energy	NYSE	16.96	EP	El Paso Corp.	NYSE	16.08	0.38	0.02
AN	AutoNation Inc.	NYSE	19.95	IPG	Interpublic Group	NYSE	11.23	0.39	0.03
AW	Allied Waste Industries	NYSE	12.71	HBAN	Huntington Bancshares	NASDAQ	19.95	0.39	0.03
LUV	Southwest Airlines	NYSE	14.78	HBAN	Huntington Bancshares	NASDAQ	19.95	0.39	0.03
DUK	Duke Energy	NYSE	19.34	CMS	CMS Energy	NYSE	17.18	0.40	0.03
JDSU	JDS Uniphase Corp.	NASDAQ	14.72	MU	Micron Technology	NYSE	11.38	0.40	0.07
AN	AutoNation Inc.	NYSE	19.95	TER	Teradyne Inc.	NYSE	15.07	0.40	0.02
SLE	Sara Lee Corp.	NYSE	16.73	EP	El Paso Corp.	NYSE	16.08	0.41	0.09
CNP	CenterPoint Energy	NYSE	17.52	EP	El Paso Corp.	NYSE	16.08	0.41	0.03
TSN	Tyson Foods	NYSE	19.06	ETFC	E*Trade Financial Corp.	NASDAQ	17.65	0.41	0.05
TER	Teradyne Inc.	NYSE	15.07	HBAN	Huntington Bancshares	NASDAQ	19.95	0.42	0.04
CMS	CMS Energy	NYSE	17.18	CNP	CenterPoint Energy	NYSE	17.52	0.43	0.06
HCBK	Hudson City Bancorp	NASDAQ	13.85	HBAN	Huntington Bancshares	NASDAQ	19.95	0.44	0.03
DUK	Duke Energy	NYSE	19.34	WIN	Windstream Corporation	NYSE	14.26	0.44	0.11
LUV	Southwest Airlines	NYSE	14.78	AN	AutoNation Inc.	NYSE	19.95	0.45	0.03
CMS	CMS Energy	NYSE	17.18	EP	El Paso Corp.	NYSE	16.08	0.45	0.03
TE	TECO Energy	NYSE	16.96	CMS	CMS Energy	NYSE	17.18	0.46	0.04
TE	TECO Energy	NYSE	16.96	DUK	Duke Energy	NYSE	19.34	0.47	0.05
AN	AutoNation Inc.	NYSE	19.95	HBAN	Huntington Bancshares	NASDAQ	19.95	0.49	0.04
TE	TECO Energy	NYSE	16.96	CNP	CenterPoint Energy	NYSE	17.52	0.51	0.04

Table A.5: Ensemble of the 50 highest correlated stock pairs from the S&P 500 index that are averagely traded between USD 10.01 and USD 20.00. The column var_{corr} refers to the variance of the correlation of a moving 30-day window.

Portfolio 1		Portfolio 2		Portfolio 3		Portfolio 4		Portfolio 5		Portfolio 6		Portfolio 7		Portfolio 8		Portfolio 9		Portfolio 10	
AGN	KEY	ADBE	LOW	ADM	LEG	ABT	K	ADI	LM	ADBE	IBM	AA	MAS	ADI	K	AAPL	HST	AA	JEC
AIG	KMB	ADSK	LTD	ADSK	LMT	ADI	LH	AEP	LUK	ADI	IGT	ADBE	MAT	AEP	KEY	ABT	HSY	ABT	KMB
ALTR	KO	AGN	MBI	AES	LOW	ADM	LLY	AET	MAS	ADP	K	AET	MEE	AET	KO	AEP	IGT	ADSK	KO
AMAT	L	AMAT	MCD	AMD	LSI	ADP	LTD	AIG	MBI	AEP	KEY	AFL	MKC	AGN	L	AES	JNJ	AEP	L
AMD	LEN	AMD	MDP	AVY	LTD	ADSK	LUK	ALTR	MDP	AES	LEN	AGN	MOLX	AMAT	LEN	AGN	JWN	AET	LLTC
APA	LLTC	APC	MHP	AYE	MBI	AES	MDP	APD	MHP	AMD	LNC	AMD	MRK	AMD	LLTC	AIG	KO	AMAT	LM
AVP	MBI	ARG	MMC	AZO	MMC	AFL	MHP	AYE	MKC	AMGN	LUK	APC	MTB	APC	LLY	AMAT	KR	AMD	LMT
AYE	MDT	BA	MSFT	BAC	MOT	AGN	MKC	BBY	MMC	APA	MAS	APH	MU	AYE	LSI	AMGN	L	AMGN	LSI
BA	MEE	BBT	MWV	BAX	MRK	AIG	MOT	BF-B	MO	APC	MBI	BEN	NEM	AZO	LUV	APA	LSI	APD	LTD
BAX	MI	BDX	NEM	BDX	MUR	AMAT	MRO	BIIB	MOLX	ARG	MEE	BMS	NI	BAC	MBI	AXP	LTD	ARG	MBI
BF-B	MRO	BHI	NKE	BEN	MYL	AMD	MWV	CAG	MOT	AVY	MHP	CA	NKE	BDX	MEE	AYE	MAT	AYE	MEE
BHI	MU	BIIB	NOVL	BF-B	NBL	APA	MYL	CAT	NBR	BBY	MI	CEG	NOC	BIIB	MHP	BA	MDP	BAC	MRK
BK	MUR	BLL	NUE	BMC	NBR	APC	NBL	CB	NEM	BCR	MMC	CELG	NU	BK	MI	BAX	MHP	BBT	MRO
CAH	MWV	CMCSA	ORCL	CA	NEM	APD	NBR	CEPH	NTRS	BEN	MO	CINF	NVLS	CAT	MRO	BBT	MI	BDX	MU
CAT	NSM	CTAS	PAYX	CAT	NI	ARG	NWL	CLX	NU	CA	MTB	COG	OMC	CLX	MWV	BBY	MOT	BEN	NSM
CB	NTRS	CVX	PBI	CEPH	NKE	AVP	ODP	CSC	OXY	CEG	NKE	CSCO	PBCT	CMA	MYL	BMC	MTB	BIIB	NTRS
CINF	OI	D	PCAR	CINF	NTRS	AXP	OMC	CTAS	PBCT	CI	NOC	CVX	PCAR	CMCSA	NBR	BMS	NBL	BLL	NWL
CLX	PBCT	DELL	PCL	CL	NVLS	BBT	PAYX	CVS	PCP	CL	NOVL	D	PCL	COST	NYT	BMY	NSC	BMC	OMC
CMA	PCL	DIS	PEP	CMI	OI	BK	PCL	CVX	PG	CMS	NTRS	DELL	PPG	CTAS	PBI	CA	NUE	BMY	ORCL
CMI	PEP	DNB	PGR	CMS	OMC	BLL	PFE	D	PGN	COST	ORCL	DOV	PPL	CVX	PEP	COG	ODP	C	PGN
COP	PFE	DOV	PHM	CNP	ORCL	BMS	PLL	DELL	PH	CSX	OXY	ECL	PSA	DD	PFE	COST	PBCT	CCL	PGR
CSX	PKI	ERTS	PKI	COG	PEP	CA	PNC	DIS	PNC	CTAS	PBI	EK	QCOM	DE	PGN	CPB	PEP	CEPH	PLL
CTAS	PLL	ESV	PLL	CPB	PG	CAG	PNW	DNB	PPL	CTL	PCP	EMC	RDC	DELL	PKI	CSX	PFE	CMA	QCOM
CTL	PPL	ETR	PNW	CSX	PGN	CAT	PPG	DV	PSA	CVS	PG	EMR	ROK	DHR	PNW	CVS	PKI	CNP	R
D	RF	EXC	POM	CVX	PNW	CCE	PPL	ED	QCOM	CVX	PH	ESV	RSH	DIS	PSA	DE	PLL	COST	RF
DE	ROK	EXPD	PPL	DD	PSA	CELG	QCOM	EMC	R	D	PKI	FDO	SHW	DOW	R	DHR	PNC	CVS	SCHW
DHR	RSH	F	ROK	DIS	RDC	CTL	R	EMR	S	DELL	PNW	FHN	SLB	DUK	RDC	DOV	PPL	DD	SLB
DIS	SEE	FHN	S	DOV	RF	DOV	RSH	EQT	SCG	DHR	R	FO	SLE	EFX	ROK	DTE	QCOM	DELL	SLM
DOV	SIAL	FISV	SCG	DTE	SEE	DV	SCG	EXC	SIAL	DIS	ROK	GD	STI	EIX	RSH	DUK	SCG	DHR	SNA
ECL	SLB	FMC	SCHW	DUK	SIAL	ECL	SHW	F	SLE	DNB	RRD	GR	STJ	EMC	SCHW	ED	SLB	DIS	STI
ESV	SLE	FO	STJ	EMC	SLB	ED	SYMC	FLS	SO	DOV	RSH	GWW	STT	EOG	SEE	EIX	SLE	DNB	SVU
ETN	STJ	FRX	STR	EQT	SO	EMR	SYY	FO	SPLS	DUK	RTN	HAL	SVU	ETN	SLB	EQT	STT	DTE	SYK
ETR	SWK	GAS	STT	ERTS	STR	ERTS	TAP	FTR	STT	ED	SCG	HBAN	SYK	EXC	SLE	ERTS	SUN	EOG	SYMC
FAST	SWN	GE	SWK	ETR	STT	FDO	TER	GLW	SVU	EFX	SLB	HES	T	FDO	SPLS	ESV	SVU	EQT	SYY
FDX	SYK	GIS	SYY	GE	SWK	FRX	TGT	GWW	SYMC	EMR	SLM	HNZ	TAP	FHN	STI	EXC	T	ESV	TAP
FHN	THC	GLW	T	GENZ	SYK	GE	TJX	HAS	TE	EQT	SNA	HON	TER	FITB	STJ	FAST	TEG	EXC	TE
FRX	TJX	GPC	TAP	GIS	TAP	GPS	TLAB	HD	TER	ERTS	STI	HOT	THC	FMC	SUN	FDO	TER	FDO	TGT
GAS	TLAB	HAL	TE	GPC	THC	GR	TSN	HPQ	THC	F	STT	HPQ	TIF	FRX	TE	FLS	TGT	FDX	TJX
GE	TROW	HAS	THC	GR	TJX	GT	TSS	HSY	TJX	FAST	SVU	HRL	TLAB	GT	TER	GAS	THC	FHN	TSN
GWW	TSN	HNZ	TIF	HAL	TMK	GWW	TXN	IFF	TLAB	FDX	TAP	HRS	TMO	HBAN	TIF	GCI	TJX	FITB	TXT
HAL	TSS	HUM	TMK	HBAN	TROW	HD	UNP	ITW	TMO	FITB	TGT	HST	TSN	HD	TJX	GD	TMK	FRX	UNM
HCP	TXN	IBM	TRV	HCP	TXN	HES	VAR	JCI	TSS	FLS	TROW	HSY	TXT	HON	TMO	GENZ	TSS	GAS	UNP
HPQ	UNH	IFF	VMC	HRB	UNM	HNZ	VMC	JCP	TXN	FTR	UNH	IGT	UNP	HOT	TRV	GIS	UNM	GCI	UTX
HST	VFC	IGT	VNO	HRL	VAR	HOT	WEC	KBH	TXT	GD	UNP	INTC	USB	INTC	TSN	GLW	UNP	GPS	WDC
HSY	VNO	IP	WMB	ITW	VZ	HRB	WHR	KEY	UNH	GENZ	WHR	JNJ	VLO	IP	TSS	GT	VNO	HAS	WHR
INTC	X	JCP	WMT	JCI	WEC	HRS	WMB	KLAC	WDC	GLW	WPO	KEY	WDC	IPG	UTX	GWW	VZ	HD	WMB
ITW	XEL	JEC	X	JCP	WPO	HST	WPO	KMB	WMT	GPS	X	LEG	WY	ITW	WAG	HCP	WAG	HRB	WPO
JNJ	XL	L	XL	KMB	XEL	HSY	X	KR	WY	GT	XEL	LLY	XL	JCI	WHR	HNZ	WDC	HRL	WY
K	XLNX	LEG	XOM	KO	XRAY	IPG	XEL	LEG	XRAY	HES	XOM	LMT	XRX	JCP	WY	HOT	WHR	IFF	XEL
KBH	XRX	LLY	XRAY	L	ZION	JPM	XL	LLTC	ZION	HRL	XRAY	LUV	ZION	JWN	XL	HRS	XRAY	INTC	XRX

Table A.6: Randomly chosen portfolios from the S&P 500 constituents (S&P symbols).

B Derivations for Credit Risk

B.1 Expansion of $\det(\mathbf{I}N + T\mathbf{S}\vec{\omega}\vec{\omega}^\dagger\mathbf{S})$

Since the matrix $\mathbf{S}\vec{\omega}\vec{\omega}^\dagger\mathbf{S}$ has rank one, the determinant can be expressed as

$$\det(\mathbf{I}N + T\mathbf{S}\vec{\omega}\vec{\omega}^\dagger\mathbf{S}) = N^K \det(\mathbf{I} + \frac{T}{N}\mathbf{S}\vec{\omega}\vec{\omega}^\dagger\mathbf{S}) \tag{B.1}$$

$$= N^K \exp\left(\ln\left(\operatorname{tr}\left(\mathbf{I} + \frac{T}{N}\mathbf{S}\vec{\omega}\vec{\omega}^\dagger\mathbf{S}\right)\right)\right) . \tag{B.2}$$

Now we express the logarithm with a power series,

$$\det(\mathbf{I}N + T\mathbf{S}\vec{\omega}\vec{\omega}^\dagger\mathbf{S}) = N^K \exp\left(\sum_{\nu=1}^{\infty} \frac{(-1)^{\nu+1}}{\nu} \left(\frac{T}{N}\operatorname{tr}(\mathbf{S}\vec{\omega}\vec{\omega}^\dagger\mathbf{S})\right)^\nu\right) \tag{B.3}$$

$$= N^K \exp\left(\sum_{\nu=1}^{\infty} \frac{(-1)^{\nu+1}}{\nu} \left(\frac{T}{N}\operatorname{tr}(\vec{\omega}^\dagger\mathbf{S}\mathbf{S}\vec{\omega})\right)^\nu\right) . \tag{B.4}$$

As $\vec{\omega}^\dagger\mathbf{S}\mathbf{S}\vec{\omega}$ is a scalar, we can remove the trace and write the the power series again as a logarithm,

$$\det(\mathbf{I}N + T\mathbf{S}\vec{\omega}\vec{\omega}^\dagger\mathbf{S}) = N^K \exp\left(\sum_{\nu=1}^{\infty} \frac{(-1)^{\nu+1}}{\nu} \left(\frac{T}{N}\vec{\omega}^\dagger\mathbf{S}\mathbf{S}\vec{\omega}\right)^\nu\right) \tag{B.5}$$

$$= N^K \exp\left(\ln(1 + \frac{T}{N}\vec{\omega}^\dagger\mathbf{S}\mathbf{S}\vec{\omega})\right) \tag{B.6}$$

$$= N^K \left(1 + \frac{T}{N}\vec{\omega}^\dagger\mathbf{S}\mathbf{S}\vec{\omega}\right) . \tag{B.7}$$

B.2 Calculation of Ω_{K-1}

We can calculate Ω_{K-1} with the help of a Gaussian integral,

$$\int d(V) \exp\left(\sum_{k=1}^{K} \frac{V_k^2}{\sigma_k^2}\right) = \int_0^\infty d\rho \rho^{K-1} \exp\left(-\rho^2\right) \int_0^\pi d\vartheta \sin(\vartheta)^{K-2} \int d\Omega_{K-2} \quad \text{(B.8)}$$

$$\prod_{k=1}^{K} \int_{-\infty}^{+\infty} dV_k \exp\left(-\frac{V_k^2}{\sigma_k^2}\right) = \frac{\Gamma(K/2)}{2} \frac{\sqrt{\pi}\Gamma((K-1)/2)}{\Gamma(K/2)} \int d\Omega_{K-2} \quad \text{(B.9)}$$

$$\int d\Omega_{K-2} = \frac{2\pi^{(K-1)/2}}{\Gamma((K-1)/1)} \prod_{k=1}^{K} \sigma_k \ . \quad \text{(B.10)}$$

B.3 $M_{1,k}(z)$ and $M_{2,k}(z)$

$$\begin{aligned}
M_{1,k}(z) = \frac{\sqrt{N}}{2\sqrt{Tz}} & \left(\left(\sqrt{Tz} \mathrm{Erf}\left(\frac{1}{4} \sqrt{\frac{N\left(-2\mu_k T + \sigma_k^2 T + 2\ln(F) - 2\ln(V_{k,0})\right)^2}{\sigma_k^2 Tz}} \right) \right.\right. \\
\times & \left(-2\mu_k T + \sigma_k^2 T + 2\ln(F) - 2\ln(V_{k,0})\right) \\
+ & \left. \sqrt{Tz\left(-2\mu_k T + \sigma_k^2 T + 2\ln(F) - 2\ln(V_{k,0})\right)^2} \right) \\
/ & \left(\sqrt{N\left(\left(-2\mu_k + \sigma_k^2\right) T + 2\ln(F) - 2\ln(V_{k,0})\right)^2} \right) \\
- & \left(\exp\left(\mu_k T - \frac{\sigma_k^2 T}{2} + \frac{\sigma_k^2 T z}{N} \right) V_{k,0} \right. \\
\times & \left(\mathrm{Erf}\left(\frac{1}{2} \sqrt{\frac{N\left(\frac{1}{2} T\left(-2\mu_k + \sigma_k^2 - \frac{4\sigma_k^2 z}{N}\right) + \ln(F) - \ln(V_{k,0})\right)^2}{\sigma_k^2 Tz}} \right) \right. \\
\times & \left(T\left(-2\mu_k N + \sigma_k^2(N - 4z)\right) + 2N(\ln(F) - \ln(V_{k,0}))\right) + \\
\times & \left.\left. \sqrt{\left(T\left(2\mu_k N - \sigma_k^2(N-4z)\right) + 2N(-\ln(F) + \ln(V_{k,0}))\right)^2} \right) \right) \\
/ & \left.\left(F \sqrt{\frac{N\left(T\left(2\mu_k N - \sigma_k^2(N-4z)\right) - 2N\ln(F) + 2N\ln(V_{k,0})\right)^2}{Tz}} \right) \right) \quad \text{(B.11)}
\end{aligned}$$

$$
\begin{aligned}
M_{2,k}(z) = \frac{1}{2\sqrt{N}} & \left(- \left(\operatorname{Erf}\left(\frac{1}{4} \sqrt{\frac{N\left(-2\mu_k T + \sigma_k^2 T + 2\ln(F) - 2\ln(V_{k,0})\right)^2}{\sigma_k^2 T z}} \right) \right.\right. \\
\times & \left. \sqrt{N\left(-2\mu_k T + \sigma_k^2 T + 2\ln(F) - 2\ln(V_{k,0})\right)^2} \right) \\
/ & \left(2\mu_k T - \sigma_k^2 T - 2\ln(F) + 2\ln(V_{k,0})\right) + \frac{\sqrt{Tz}}{F^2} \\
\times & \left(- \left(\exp\left(T\left(2\mu_k - \frac{\sigma_k^2 (N-4z)}{N} \right) \right) V_{k,0}^2 \right.\right. \\
\times & \operatorname{Erf}\left(\frac{1}{2} \sqrt{\frac{N\left(\frac{1}{2}T\left(-2\mu_k + \sigma_k^2 - \frac{8\sigma_k^2 z}{N}\right) + \ln(F) - \ln(V_{k,0})\right)^2}{\sigma_k^2 T z}} \right) \\
\times & \left. \sqrt{\frac{N\left(T\left(2\mu_k N - \sigma_k^2 (N-8z)\right) + 2N(-\ln(F) + \ln(V_{k,0}))\right)^2}{Tz}} \right) \\
/ & \left(T\left(2\mu_k N - \sigma_k^2(N-8z)\right) + 2N(-\ln(F) + \ln(V_{k,0}))\right) + \exp\left(-\sigma_k^2 T\right) \\
\times & \left(\frac{\exp\left(\sigma_k^2 T\right) F^2 - 2\exp\left(\mu_k T + \frac{\sigma_k^2 T(N+2z)}{2N}\right) F V_{k,0} + \exp\left(2T\left(\mu_k + \frac{2\sigma_k^2 z}{N}\right)\right) V_{k,0}^2}{\sqrt{\frac{Tz}{N}}} \right. \\
+ & \left(2\exp\left(\mu_k T + \frac{\sigma_k^2 T(N+2z)}{2N} \right) \right. \\
\times & F V_{k,0} \operatorname{Erf}\left(\frac{1}{2} \sqrt{\frac{N\left(\frac{1}{2}T\left(-2\mu_k + \sigma_k^2 - \frac{4\sigma_k^2 z}{N}\right) + \ln(F) - \ln(V_{k,0})\right)^2}{\sigma_k^2 T z}} \right) \\
\times & \left. \sqrt{\frac{N\left(T\left(2\mu_k N - \sigma_k^2(N-4z)\right) + 2N(-\ln(F) + \ln(V_{k,0}))\right)^2}{Tz}} \right) \\
/ & \left. \left. \left. \left(T\left(2\mu_k N - \sigma_k^2(N-4z)\right) + 2N(-\ln(F) + \ln(V_{k,0}))\right) \right) \right) \right)
\end{aligned}
\tag{B.12}
$$

Bibliography

[1] M. C. Münnix, R. Schäfer, and T. Guhr, "Compensating Asynchrony Effects in the Calculation of Financial Correlations," *Physica A*, vol. 389, no. 4, pp. 767–779, 2010.

[2] M. C. Münnix, R. Schäfer, and T. Guhr, "Impact of the Tick-size on Financial Returns and Correlations," *Physica A*, vol. 389, no. 21, pp. 4828–4843, 2010.

[3] M. C. Münnix, R. Schäfer, and O. Grothe, "Estimating Correlation and Covariance Matrices by Weighting of Market Similarity," *Quantitative Finance*, 2011. accepted for publication, arXiv:1006.5847.

[4] M. C. Münnix, R. Schäfer, and T. Guhr, "Statistical causes for the Epps effect in microstructure noise," *Journal of Theoretical and Applied Finance*, 2011. accepted for publication, arXiv:1009.6157.

[5] M. C. Münnix and R. Schäfer, "A Copula Approach on the Dynamics of Statistical Dependencies in the US Stock Market," *Physica A*, 2011. accepted for publication, arXiv:1102.1099.

[6] M. C. Münnix, T. Shimada, R. Schäfer, F. Levraz, T. Seligman, T. Guhr, and H. Stanley, "Identifying the States of a financial Market." working paper, 2011.

[7] M. C. Münnix, R. Schäfer, and T. Guhr, "A Random Matrix Approach on Credit Risk." working paper, arXiv:1102.3900, 2011.

[8] J. Voit, *The Statistical Mechanics of Financial Markets.* Heidelberg: Springer, 2001.

[9] R. N. Mantegna and H. E. Stanley, *An introduction to econophysics: correlations and complexity in finance.* Cambridge: Cambridge University Press, 2000.

[10] V. Plerou, P. Gopikrishnan, B. Rosenow, L. A. N. Amaral, T. Guhr, and H. E. Stanley, "Random matrix approach to cross correlations in financial data," *Physical Review E*, vol. 65, p. 066126, Jun 2002.

[11] L. Laloux, P. Cizeau, J.-P. Bouchaud, and M. Potters, "Noise dressing of financial correlation matrices," *Physical Review Letters*, vol. 83, pp. 1467–1470, Aug 1999.

[12] E. F. Fama, "The behavior of stock-market prices," *The Journal of Business*, vol. 38, no. 1, pp. 34–105, 1965.

[13] E. F. Fama, "Efficient capital markets: A review of theory and empirical work," *The Journal of Finance*, vol. 25, no. 2, pp. 383–417, 1970.

[14] K. Pearson, "On the criterion that a given system of deviations from the probable in the case of a correlated system of variables is such that it can be reasonably supposed to hove arisen from random sampling," *Philosophical Magazine*, no. 50, pp. 157–172, 1900.

[15] J. Bouchaud and M. Potters, *Theory of Financial Risks.* Cambridge: Cambridge University Press, 2000.

[16] J. B. Perrin, "Mouvement brownien et réalité moléculaire," *Annales de Chimie et de Physique*, vol. 8, no. 18, pp. 4–114, 1909.

[17] L. Bachelier, *The theory of speculation.* PhD thesis, Princeton University, 1900.

[18] A. Einstein, "Über die von der molekularkinetischen Theorie der Wärme geforderte Bewegung von in ruhenden Flüssigkeiten suspendierten Teilchen," *Annalen der Physik*, vol. 322, no. 8, pp. 549–560, 1905.

[19] F. Black and M. S. Scholes, "The pricing of options and corporate liabilities," *Journal of Political Economy*, vol. 81, no. 3, pp. 637–54, 1973.

[20] P. Gopikrishnan, V. Plerou, L. A. Nunes Amaral, M. Meyer, and H. E. Stanley, "Scaling of the distribution of fluctuations of financial market indices," *Physical Review E*, vol. 60, pp. 5305–5316, Nov 1999.

[21] V. Plerou, P. Gopikrishnan, L. A. Nunes Amaral, M. Meyer, and H. E. Stanley, "Scaling of the distribution of price fluctuations of individual companies," *Phys. Rev. E*, vol. 60, pp. 6519–6529, Dec 1999.

[22] B. Mandelbrot, "The Variation of Certain Speculative Prices," *Journal of Business*, vol. 36, p. 394, 1963.

[23] cont01, "Empirical properties of asset returns: stylized facts and statistical issues," *Quantitative Finance*, vol. 1, pp. 223–236, 2001.

[24] A. M. Sengupta and P. P. Mitra, "Distributions of singular values for some random matrices," *Phys. Rev. E*, vol. 60, no. 3, pp. 3389–3392, 1999.

[25] F. J. Dyson, "Distribution of eigenvalues for a class of real symmetric matrices," *Revista Mexicana de Física*, vol. 20, p. 231, 1971.

[26] T. Guhr and B. Kälber, "A new method to estimate the noise in financial correlation matrices," *Journal of Physics A: Mathematical and General*, vol. 36, no. 12, pp. 3009–3032, 2003.

[27] R. Schäfer, N. Nilsson, and T. Guhr, "Power mapping with dynamical adjustment for improved portfolio optimization," *Quantitative Finance*, 2009.

[28] J. Schäfer and K. Strimmer, "A shrinkage approach to large-scale covariance matrix estimation and implications for functional genomics.," *Statistical applications in genetics and molecular biology*, vol. 4, 2005.

[29] O. Ledoit and M. Wolf, "Improved estimation of the covariance matrix of stock returns with an application to portfolio selection," *Journal of Empirical Finance*, vol. 10, no. 5, pp. 603–621, 2003.

[30] S. Lee, "Forbes.com: Formula From Hell - The Gaussian copula and the market meltdown," 2009.

[31] R. Schäfer, M. Sjölin, A. Sundin, M. Wolanski, and T. Guhr, "Credit risk–a structural model with jumps and correlations," *Physica A*, vol. 383, no. 2, pp. 533–569, 2007.

[32] R. N. Mantegna and H. E. Stanley, "Stock market dynamics and turbulence: parallel analysis of fluctuation phenomena," *Physica A*, vol. 239, no. 1-3, pp. 255–266, 1997.

[33] C. Borghesi, M. Marsili, and S. Miccichè, "Emergence of time-horizon invariant correlation structure in financial returns by subtraction of the market mode," *Physical Review E*, vol. 76, no. 2, p. 026104, 2007.

[34] T. F. Cooley and V. Quadrini, "Financial markets and firm dynamics," *The American Economic Review*, vol. 91, no. 5, pp. 1286–1310, 2001.

[35] D. Pelletier, "Regime switching for dynamic correlations," *Journal of Econometrics*, vol. 131, no. 1-2, pp. 445–473, 2006.

[36] X. E. Xu, P. Chen, and C. Wu, "Time and dynamic volume-volatility relation," *Journal of Banking & Finance*, vol. 30, no. 5, pp. 1535–1558, 2006.

[37] M. King and S. Wadhwani, "Transmission of volatility between stock markets," *The Review of Financial Studies*, vol. 3, no. 1, pp. 5–33, 1990.

[38] S. Lee, "The stability of the co-movements between real estate returns in the uk," *Journal of Property Investment & Finance*, vol. 24, no. 5, pp. 434–442, 2006.

[39] V. Plerou, P. Gopikrishnan, and E. Stanley, "Econophysics: Two-phase behaviour of financial markets," *Nature*, vol. 421, no. 130, 2003.

[40] B. Rosenow, P. Gopikrishnan, V. Plerou, and H. E. Stanley, "Dynamics of cross-correlations in the stock market," *Physica A*, vol. 324, no. 1-2, pp. 241–246, 2003. Proceedings of the International Econophysics Conference.

[41] H. Tastan, "Estimating time-varying conditional correlations between stock and foreign exchange markets," *Physica A*, vol. 360, no. 2, pp. 445–458, 2006.

[42] S. Drozdz, J. Kwapien, F. GrÂ¸mmer, F. Ruf, and J. Speth, "Quantifying the dynamics of financial correlations," *Physica A*, vol. 299, no. 1-2, pp. 144–153, 2001.

[43] "New York Stock Exchange, Monthly TAQ DVD, License #400570," 2007.

[44] R. Schäfer and T. Guhr, "Local normalization: Uncovering correlations in non-stationary financial time series," *Physica A*, vol. 389, no. 18, pp. 3856–3865, 2010.

[45] "What The Market Is Telling Us," *Bloomberg Businessweek*, Mar 12 2007. Cover story.

[46] "A.I.G. Sells $39.3 Billion in Assets to N.Y. Fed's Fund'," *The New York Times*, Dec 16 2008.

[47] "A.I.G. Moves Ahead With Asset Disposals, Report Says," *The New York Times*, Jan 13 2009.

[48] J. B. MacQueen, "Some methods for classification and analysis of multivariate observations," in *Proc. of the fifth Berkeley Symposium on Mathematical Statistics and Probability* (L. M. L. Cam and J. Neyman, eds.), vol. 1, pp. 281–297, University of California Press, 1967.

[49] V. Faber, "Clustering and the continuous k-means algorithm," *Los Alamos Science*, vol. 22, pp. 138–144, 1994.

[50] N. K. Tanaka, T. Awasaki, T. Shimada, and K. Ito, "Integration of chemosensory pathways in the drosophila second-order olfactory centers," *Current biology*, vol. 14, no. 6, pp. 449–457, 2004.

[51] "http://www.standardandpoors.com/indices/gics/en/us."

[52] H. M. Markowitz, "Portfolio selection," *The Journal of Finance*, vol. 7, no. 1, pp. 77–91, 1952.

[53] H. M. Markowitz, *Portfolio Selection: Efficient Diversification of Investments*. John Wiley & Sons, Inc., New York, 1959.

[54] E. J. Elton and M. J. Gruber, "Modern portfolio theory, 1950 to date," *Journal of Banking & Finance*, vol. 21, no. 11-12, pp. 1743–1759, 1997.

[55] E. F. Fama, "Portfolio analysis in a stable paretian market," *Management Science*, vol. 11, no. 3, pp. 404–419, 1965.

[56] E. J. Elton and M. J. Gruber, "Portfolio theory when investment relatives are lognormally distributed," *The Journal of Finance*, vol. 29, no. 4, pp. 1265–1273, 1974.

[57] A. Kraus and R. H. Litzenberger, "Skewness preference and the valuation of risk assets," *The Journal of Finance*, vol. 31, no. 4, pp. 1085–1100, 1976.

[58] C. F. Lee, "Functional form, skewness effect, and the risk-return relationship," *The Journal of Financial and Quantitative Analysis*, vol. 12, no. 1, pp. 55–72, 1977.

[59] R. Engle and R. Colacito, "Testing and valuing dynamic correlations for asset allocation," *Journal of Business & Economic Statistics*, vol. 24, pp. 238–253, April 2006.

[60] T. J. Brailsford and R. W. Faff, "An evaluation of volatility forecasting techniques," *Journal of Banking & Finance*, vol. 20, no. 3, pp. 419–438, 1996.

[61] G. A. Vasilellis and N. Meade, "Forecasting volatility for portfolio selection," *Journal of Business Finance & Accounting*, vol. 23, no. 1, pp. 125–143, 1996.

[62] R. D. Harris and F. Yilmaz, "Estimation of the conditional variance-covariance matrix of returns using the intraday range," *International Journal of Forecasting*, vol. 26, no. 1, pp. 180–194, 2010. Special Section: European Election Forecasting.

[63] A. P. Nawroth, R. Friedrich, and J. Peinke, "Multi-scale description and prediction of financial time series," *New Journal of Physics*, vol. 12, no. 8, p. 083021, 2010.

[64] A. P. Nawroth and J. Peinke, "Multiscale reconstruction of time series," *Physics Letters A*, vol. 360, no. 2, pp. 234–237, 2006.

[65] J. G. D. Gooijer and R. J. Hyndman, "25 years of time series forecasting," *International Journal of Forecasting*, vol. 22, no. 3, pp. 443 – 473, 2006. Twenty five years of forecasting.

[66] J. Longerstaey and M. Spencer, "RiskMetrics Technical Document," 4th edition, J.P.Morgan/Reuters, New York, 1996.

[67] S. Lee and S. Stevenson, "Time weighted portfolio optimisation," *Journal of Property Investment & Finance*, vol. 21, no. 3, pp. 233–249, 2003.

[68] O. Ledoit and M. Wolf, "Honey, I Shrunk the Sample Covariance Matrix," *The Journal of Portfolio Management*, vol. 30, no. 4, pp. 110–119, 2004.

[69] T. Bollerslev, "Generalized autoregressive conditional heteroskedasticity," *Journal of Econometrics*, vol. 31, no. 3, pp. 307–327, 1986.

[70] J. Voit, *The Statistical Mechanics of Financial Markets*, ch. 8. Microscopic Market Models. Heidelberg: Springer, 2001.

[71] F. Bandi, J. Russell, and Y. Zhu, "Using high-frequency data in dynamic portfolio choice," *Econometric Reviews*, vol. 27, no. 1–3, pp. 163–198, 2008.

[72] M. Kritzman, S. Page, and D. Turkington, "In defense of optimization: The fallacy of 1/n," *Financial Analysts Journal*, vol. 66, no. 2, pp. 31–39, 2010.

[73] V. DeMiguel, L. Garlappi, and R. Uppal, "Optimal Versus Naive Diversification: How Inefficient is the 1/N Portfolio Strategy?," *The Review of Financial Studies*, vol. 22, no. 5, pp. 1915–1953, 2009.

[74] C. Spearman, "The proof and measurement of association between two things," *The American Journal of Psychology*, vol. 100, no. 3/4, pp. 441–471, 1987.

[75] W. F. Sharpe, "Capital asset prices: A theory of market equilibrium under conditions of risk," *The Journal of Finance*, vol. 19, no. 3, pp. 425–442, 1964.

[76] J. D. Noh, "Model for correlations in stock markets," *Physical Review E*, vol. 61, pp. 5981–5982, May 2000.

[77] A. Sklar, "Fonctions de répartition à n dimensions et leurs marges," *Publ. Inst. Statist. Univ. Paris*, vol. 8, pp. 229–231, 1959.

[78] A. Sklar, "Random variables, joint distribution functions, and copulas," *Kybernetika*, vol. 9, no. 6, pp. 449–460, 1973.

[79] R. B. Nelson, *An Introduction to Copulas*. Springer, 1998.

[80] V. Fernandez, “Copula-based measures of dependence structure in assets returns,” *Physica A*, vol. 387, no. 14, pp. 3615–3628, 2008.

[81] V. Chavez-Demoulin, P. Embrechts, and J. Neslehov, “Quantitative models for operational risk: Extremes, dependence and aggregation,” *Journal of Banking & Finance*, vol. 30, no. 10, pp. 2635–2658, 2006.

[82] J. V. Rosenberg and T. Schuermann, “A general approach to integrated risk management with skewed, fat-tailed risks,” *Journal of Financial Economics*, vol. 79, no. 3, pp. 569–614, 2006.

[83] Y. Malevergne and D. Sornette, “Testing the gaussian copula hypothesis for financial assets dependences,” *Quantitative Finance*, vol. 3, no. 4, pp. 231–250, 2003.

[84] G. Frahm, M. Junker, and R. Schmidt, “Estimating the tail-dependence coefficient: Properties and pitfalls,” *Insurance: Mathematics and Economics*, vol. 37, no. 1, pp. 80–100, 2005. Papers presented at the DeMoSTAFI Conference, QuÈbec, 20-22 May 2004.

[85] R. Schmidt and U. Stadtmüller, “Non-parametric estimation of tail dependence,” *Scandinavian Journal of Statistics*, vol. 33, no. 2, pp. 307–335, 2006.

[86] J. E. Heffernan, “A directory of coefficients of tail dependence,” *Extremes*, vol. 3, pp. 279–290, 2000. 10.1023/A:1011459127975.

[87] A. M. Petersen, S. H. F. Wang, and H. E. Stanley, “Quantitative law describing market dynamics before and after interest-rate change,” vol. 81, p. 066121, 2010.

[88] L. Kullmann, J. Kertész, and K. Kaski, “Time-dependent cross-correlations between different stock returns: A directed network of influence,” *Physical Review E*, vol. 66, p. 026125, Aug 2002.

[89] T. W. Epps, “Comovements in stock prices in the very short run,” *Journal of the American Statistical Association*, vol. 74, no. 366, pp. 291–298, 1979.

[90] G. Bonanno, F. Lillo, and R. N. Mantegna, “High-frequency cross-correlation in a set of stocks,” *Quantitative Finance*, vol. 1, no. 1, pp. 96–104, 2001.

[91] J. Kwapien, S. Drozdz, and J. Speth, “Time scales involved in emergent market coherence,” *Physica A*, vol. 337, no. 1-2, pp. 231–242, 2004.

[92] M. Tumminello, T. D. Matteo, T. Aste, and R. Mantegna, “Correlation based networks of equity returns sampled at different time horizons,” *The European Physical Journal B*, vol. 55, pp. 209–217, jan 2007.

[93] A. A. Zebedee and M. Kasch-Haroutounian, "A closer look at co-movements among stock returns," *Journal of Economics and Business*, vol. 61, no. 4, pp. 279–294, 2009.

[94] M. Lundin, M. Dacorogna, and U. A. Müller, *Financial Markets Tick By Tick*, ch. Correlation of High Frequency Financial Time Series. Wiley, 1998.

[95] J. Muthuswamy, S. Sarkar, A. Low, and E. Terry, "Time variation in the correlation structure of exchange rates: high-frequency analyses," *Journal of Futures Markets*, vol. 21, no. 2, pp. 127–144, 2001.

[96] J.-P. Onnela, A. Chakraborti, K. Kaski, J. Kertész, and A. Kanto, "Dynamics of market correlations: Taxonomy and portfolio analysis," *Physical Review E*, vol. 68, p. 056110, Nov 2003.

[97] T. Hayashi and N. Yoshida, "On covariance estimation of non-synchronously observed diffusion processes," *Bernoulli*, vol. 11, no. 2, pp. 359–379, 2005.

[98] V. Voev and A. Lunde, "Integrated Covariance Estimation using High-frequency Data in the Presence of Noise," *Journal of Financial Econometrics*, vol. 5, no. 1, pp. 68–104, 2007.

[99] J. E. Griffin and R. C. Oomen, "Covariance measurement in the presence of non-synchronous trading and market microstructure noise," *Journal of Econometrics*, vol. 160, no. 1, pp. 58–68, 2011.

[100] B. Tóth and J. Kertész, "The epps effect revisited," *Quantitative Finance*, vol. 9, no. 7, pp. 793–802, 2009.

[101] R. A. Edelson and J. H. Krolik, "The discrete correlation function - A new method for analyzing unevenly sampled variability data," *Astrophysical Journal*, vol. 333, pp. 646–659, Oct. 1988.

[102] F. Corsi and F. Audrino, "Realized Correlation Tick-By-Tick," *SSRN eLibrary*, 2007.

[103] L. Zhang, "Estimating covariation: Epps effect, microstructure noise," *Journal of Econometrics*, vol. 160, no. 1, pp. 33–47, 2011.

[104] O. E. Barndorff-Nielsen, P. R. Hansen, A. Lunde, and N. Shephard, "Multivariate realised kernels: consistent positive semi-definite estimators of the covariation of equity prices with noise and non-synchronous trading," Economics Papers 2008-W10, Economics Group, Nuffield College, University of Oxford, 2008.

[105] O. E. Barndorff-Nielsen and N. Shephard, "Econometric analysis of realised covariation: High frequency covariance, regression and correlation in financial economics," Economics Papers 2002-W13, Economics Group, Nuffield College, University of Oxford, 2001.

[106] R. Renò, "A closer look at the epps-effect," *International Journal of Theoretical and Applied Finance*, vol. 6, pp. 87–102, February 2003.

[107] M. Martens and S.-H. Poon, "Returns synchronization and daily correlation dynamics between international stock markets," *Journal of Banking & Finance*, vol. 25, no. 10, pp. 1805–1827, 2001.

[108] United States Securities and Exchange Commission (SEC), "Order directing the exchanges and nasd to submit a decimalization implementation plan," 2000.

[109] M.-C. Beaulieu, S. K. Ebrahim, and I. G. Morgan, "Does tick size influence price discovery? Evidence from the Toronto Stock Exchange," *Journal of Futures Markets*, vol. 23, no. 1, pp. 49–66, 2003.

[110] R. D. Huang and H. R. Stoll, "Tick size, bid-ask spreads, and market structure," *Journal of Financial and Quantitative Analysis*, vol. 36, no. 04, pp. 503–522, 2001.

[111] D. Bourghelle and F. Declerck, "Why markets should not necessarily reduce the tick size," *Journal of Banking & Finance*, vol. 28, no. 2, pp. 373–398, 2004.

[112] K. H. Chung and C. Chuwonganant, "Tick Size and Quote Revisions on the NYSE," *SSRN eLibrary*, 2001.

[113] L. Harris, "Stock price clustering and discreteness," *Review of Financial Studies*, vol. 4, no. 3, pp. 389–415, 1991.

[114] J.-P. Onnela, J. Töyli, and K. Kaski, "Tick size and stock returns," *Physica A*, vol. 388, no. 4, pp. 441–454, 2009.

[115] S. Kozicki and B. Hoffman, "Rounding error: A distorting influence on index data," *Journal of Money, Credit and Banking*, vol. 36, no. 3, pp. 319–38, 2004.

[116] L. Lanz Doran, M. A. Goldstein, E. V. Golubeva, and E. N. Hughson, "Tick Size and Adverse Selection: Spurious Effects Arising from Serial Correlation," *SSRN eLibrary*, 2009.

[117] G. G. Szpiro, "Tick size, the compass rose and market nanostructure," *Journal of Banking & Finance*, vol. 22, no. 12, pp. 1559–1569, 1998.

[118] J. Voit, "From brownian motion to operational risk: Statistical physics and financial markets," *Physica A*, vol. 321, no. 1-2, pp. 286–299, 2003.

[119] R. F. Engle, "Autoregressive conditional heteroscedasticity with estimates of the variance of united kingdom inflation," *Econometrica*, vol. 50, no. 4, pp. 987–1007, 1982.

[120] R. F. Engle, "Garch 101: The use of arch/garch models in applied econometrics," *Journal of Economic Perspectives*, vol. 15, pp. 157–168, Fall 2001.

[121] P. R. Hansen and A. Lunde, "Realized variance and market microstructure noise," *Journal of Business & Economic Statistics*, vol. 24, pp. 127–161, 2006.

[122] J.-P. Bouchaud, Y. Gefen, M. Potters, and M. Wyart, "Fluctuations and response in financial markets: the subtle nature of 'random' price changes," *Quantitative Finance*, vol. 4, no. 2, pp. 176–190, 2004.

[123] C. Bluhm, L. Overbeck, and C. Wagner, *An Introduction to Credit Risk Modeling.* CRC Press, Taylor and Francis, 2002.

[124] T. R. Bielecki and M. Rutkowski, *Credit Risk: Modeling, Valuation and Hedging.* Springer, 2005.

[125] D. Duffie and K. J. Singleton, *Credit Risk: Pricing, Measurement, and Management.* Princeton University Press, 2003.

[126] D. Lando, *Credit Risk Modeling: Theory and Applications.* Princeton University Press, 2004.

[127] A. J. McNeil, R. Frey, and P. Embrechts, *Quantitative Risk Management: Concepts, Techniques, and Tools.* Princeton University Press, 2005.

[128] J. Bessis, *Risk Management in Banking.* Wiley, 2001.

[129] A. Bangia, F. X. Diebold, A. Kronimus, C. Schagen, and T. Schuermann, "Ratings migration and the business cycle, with application to credit portfolio stress testing," *Journal of Banking & Finance*, vol. 26, no. 2-3, pp. 445–474, 2002.

[130] E. I. Altman, "The importance and subtlety of credit rating migration," *Journal of Banking & Finance*, vol. 22, no. 10-11, pp. 1231–1247, 1998.

[131] K. Giesecke, *Credit Risk: Models and Management*, vol. 2, ch. Credit Risk Modeling and Valuation: An Introduction, pp. 487–525. Risk Books, 2004.

[132] R. C. Merton, "On the Pricing of Corporate Debt: The Risk Structure of Interest Rates," *Journal of Finance*, vol. 29, no. 2, pp. 449–470, 1974.

[133] R. A. Jarrow and S. M. Turnbull, "Pricing derivatives on financial securities subject to default risk," *Journal of Finance*, vol. 50, pp. 53–86, 1995.

[134] R. A. Jarrow, D. Lando, and S. M. Turnbull, "A markov model for the term structure of credit risk spreads," *Review of Financial Studies*, vol. 10, no. 2, pp. 481–523, 1997.

[135] D. Duffie and K. Singleton, "Modeling the term structure of defaultable bonds," *Review of Financial Studies*, vol. 12, pp. 687–720, 1999.

[136] J. C. Hull and A. White, "Valuing credit default swaps I: No counterparty default risk," *Journal of Derivatives*, vol. 8, no. 1, pp. 29–40, 2000.

[137] P. J. Schönbucher, *Credit Derivatives Pricing Models*. New Jersey: John Wiley & Sons, 2003.

[138] F. Black and J. C. Cox, "Valuing Corporate Securities: Some Effects of Bond Indenture Provisions," *Journal of Finance*, vol. 31, no. 2, pp. 351–367, 1976.

[139] Y. A. Katz and N. V. Shokhirev, "Default risk modeling beyond the first-passage approximation: Extended black-cox model," *Physical Review E*, vol. 82, no. 1, p. 016116, 2010.

[140] A. Asvanunt and A. Staal, *The Corporate Default Probability model in Barclays Capital POINT platform (POINT CDP)*. Portfolio Modeling, Barclays Capital, April 2009.

[141] A. Asvanunt and A. Staal, *The POINT Conditional Recovery Rate (CRR) Model*. Portfolio Modeling, Barclays Capital, August 2009.

[142] B. Rosenow and R. Weissbach, "Modelling correlations in credit portfolio risk," *Journal of Risk Management in Financial Institutions*, vol. 3, no. 1, pp. 16–30, 2009.

[143] G. M. Gupton, C. C. Finger, and M. Bhatia, "CreditMetrics – Technical Document," tech. rep., Morgan Guaranty Trust Company, 1997.

[144] "Credit Portfolio View, Approach Document und User's Manual," tech. rep., McKinsey & Company, 1998.

[145] "Credit Risk+: A Credit Risk Management Framework," tech. rep., Credit Suisse First Boston (CSFB), 1997.

[146] S. M. Ross, *Introduction to Probability Models, Ninth Edition*, ch. 10.3.2, pp. 631–632. Academic Press, 2007.

[147] F. W. J. Olver, *Asymptotics and Special Functions*, pp. 37–38. Academic Press, 1974.

[148] G. N. Watson, *A Treatise on the Theory of Bessel Functions*, p. 183. Cambridge University Press, 1944.

[149] S. Wolfram, *Mathematica: A System for Doing Mathematics by Computer.* Addison-Wesley Longman Publishing, 1988.

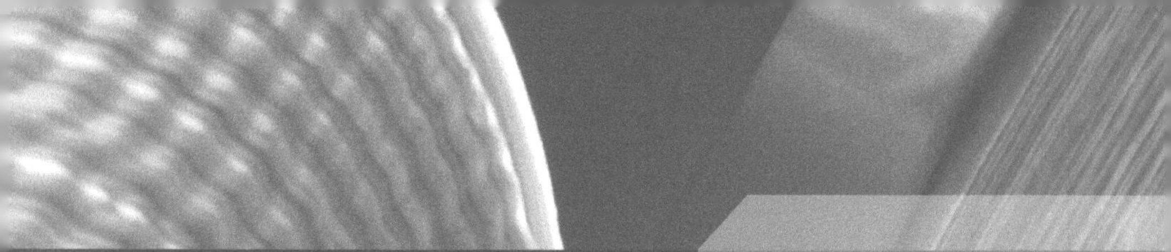

MIX
Papier aus verantwortungsvollen Quellen
Paper from responsible sources
FSC® C105338

If you have any concerns about our products,
you can contact us on
ProductSafety@springernature.com

In case Publisher is established outside the EU,
the EU authorized representative is:
Springer Nature Customer Service Center GmbH
Europaplatz 3, 69115 Heidelberg, Germany

Printed by Libri Plureos GmbH
in Hamburg, Germany